SOLAR ENERGY
IN BUILDINGS

SOLAR ENERGY IN BUILDINGS

Charles Chauliaguet
Ecole Nationale Supérieure des Beaux-Arts, Paris

Pierre Baratsabal
Architect

Jean-Pierre Batellier
Ecole Spéciale des Travaux Publics, Paris

Translated by
Joan McMullan

A Wiley–Interscience Publication

JOHN WILEY & SONS
Chichester · New York · Brisbane · Toronto

First published under Charles Chauliaguet L'Énergie Solaire Dans
Le Bâtiment by Editions Eyrolles-Paris.

© 1977 Eyrolles

English translation Copyright © 1979, by John Wiley & Sons, Ltd.

Library of Congress Cataloging in Publication Data:

Chauliaguet, Charles.
    Solar energy in building.

    Translation of L'énergie solaire dans le bâtiment.
    'A Wiley—Interscience publication.'
    Bibliography: p.
    1. Solar heating. 2. Solar air conditioning.
3. Solar houses. I. Baratsabal, Pierre, joint author.
II. Batellier, Jean Pierre, joint author. III. Title.
TH7413.C4513   697'.78   78—27031
ISBN 0 471 27570 0

Typeset in IBM Press Roman
by Preface Ltd, Salisbury, Wilts.
and printed by Unwin Brothers Ltd,
The Gresham Press, Old Woking, Surrey

The authors would like to thank Dr. M. Vigneron for his advice, friendly collaboration and the comments which he has kindly given during this study.

Bibliography

CHAPTER 2   French Meteorological Atlas.
Booklet No. 1, 'Mesures du rayonnement', A.F.E.D.E.S.
CHAPTER 3   Booklets dated 1961, C.S.T.B.
'Les déperditions des locaux d'habitation', D.T.U.
CHAPTER 4   The Congress of Rome, 1961, organized by the United Nations.
'The Sun in the Service of Man', U.N.E.S.C.O. Conference, 1973.
CHAPTER 5   I.S.E.S. Conference, Los Angeles, August, 1975.
CHAPTER 6   E.D.F. literature.
Philips literature.
CHAPTER 7   A.F.E.D.E.S. Conference, Lyons, April, 1976.

Journals: Solar Energy (published by the International Solar Energy Society).
La Comples (published half-yearly).

These publications can be consulted in the library of the Institut Français des Combustibles et de l'Energie, rue Henri–Heine, 75016 Paris.

List of Principal Symbols

G_{OH}, G_H	Total radiation on a horizontal surface under clear sky, under average sky (W m^{-2})
I_0	Direct radiation above the atmosphere (W m^{-2})
I_0^*, I^*	Direct radiation received at the ground under clear sky, under average sky (W m^{-2})
D_{OH}, D_H	Diffuse radiation on a horizontal plane under clear sky, under average sky (W m^{-2})
S_{OH}, S_H	Direct radiation component along the normal to the horizontal plane under clear sky, under average sky (W m^{-2})
D_0, D	Diffuse radiation on an inclined surface under clear sky, under average sky
α	Albedo
σ	Insolation factor
Q	Total energy received per day (Wh m^{-2} or kWh m^{-2})
h, h_{max}	Altitude of the sun, maximum altitude
ϕ	Latitude
δ	Solar declination
HA	Hour angle
a	Solar azimuth
α	Angle of a surface with respect to the south
G	Coefficient of volume losses (W m^{-3} °C^{-1})
D	Degree-days
k	Conductance of total losses in a collector
Q_a	Power absorbed in a collector (W m^{-2})
Q_p	Power dissipated or lost (W m^{-2})
Q_u	Power collected or useable (W m^{-2})
$\dot{m}C_p$	Mass flow of fluid in a collector (1 h^{-1} m^{-2})
η	Instantaneous efficiency of a collector
F_R	Efficiency factor for a collector
C_R	Concentration factor

α	Absorption coefficient
τ	Transmission coefficient
COP	Coefficient of heat pump performance

(NOTE: 1 kWh = 860 kcal)

Table of Contents

Preface to the English Edition

There are now a number of publications on the subject of solar energy. Some of these are highly technical and some are at a level which could only be termed 'popular', but very few try to bridge the gap between the theory and the practice. In addition, most of the published material is relevant only to low latitude countries with high insolation or to the United States of America. Little is available regarding more northerly regions. One must always remember that New York is at the same latitude as Madrid, and that Glasgow relates to the middle of Hudson's Bay. In Northern Europe, the climate is very different from that in most of the areas for which solar engineering studies are made.

If one wishes to make a start in bridging the gap between pure science, architecture, and building services, where better to start than in France. The French have had a long interest in solar power, dating back to Descartes and before, and they have a record of developing alternative energy systems ahead of the rest of the world. To illustrate this, one has only to think of the tidal power station on La Rance at St. Malo, and the vast one megawatt solar furnace in the Pyrenees at Odeillo, which can produce temperatures up to 3800 K. Low grade geothermal energy is also being widely developed, and much thought and experiment is being devoted to the problems associated with using solar energy intelligently in a country that stretches from the Mediterranean to the North Sea.

The authors of this book have tried to bring to the architect or to the buildings services engineer some of the background scientific information that is required before solar house designs can be properly assessed, and they do so in a manner that is readily understandable. They also introduce some considerations that are frequently overlooked, and include ways of assessing the viability of an individual site and the criteria to be adopted in deciding if a project is economically worth while. Examples are drawn from around the world, but the most detailed discussions are given on projects in France itself. This does not detract from their general applicability but increases it for those of us who live in continental Europe and in similar latitudes. This book will make a valuable addition to the library of all those who are interested in the application of solar energy in buildings.

Introduction

In France today buildings using solar energy for heating and the production of hot water can be counted in dozens, but by the 1980s there will be several thousand such buildings.

The *Délgation aux Energies nouvelles* is co-ordinating various projects of a public nature (schools, swimming pools, a crèche in Corsica, a tax office in Salon-de-Provence, etc.). These projects will permit system testing in several regions and furthermore will foster public understanding of the energy potential presented by solar radiation.

The *Ministère de l'Equipement,* through the intermediary body H.O.T., is sponsoring several ideas and is providing the opportunity of testing them.

A.F.E.D.E.S. (*Association pour le développement de l'énergie solaire*) and the French branch of C.O.M.P.L.E.S. (Coopération méditerranéenne pour *l'énergie solaire*) have brought about the present development of solar energy in France through the research of their members.

The aim of this book is to take stock of an area which is rapidly developing reliable techniques and is on the road to industrialization.

1
Solar Architecture

1.1 HISTORICAL REVIEW AND BIOENERGETIC FACTORS

1.1.1 Introduction

The main aim of building design is the creation of an artifical microclimate which satisfies contemporary definitions of thermal comfort: air speed and temperature , radiation, humidity.

This aim becomes particularly apparent in a study of successive dwelling forms undertaken in these terms, but a reduction to this single definition would not be sufficient. Current descriptive methods take into account many other factors (defensive, non-climatic, environmental, historical, etc.).

The study over a period of time of the microclimatic characteristics of habitat emphasizes the 'economy of resources', the method of selection of technological acquisitions. Thus, solar solutions are primarily applicable only in regions of high insolation levels, and it is in dry and cloudfree regions that the most elaborate systems will be developed. This over-simplification does not claim to define a climatic zone for the utilization of solar energy, but does define one in which to site the most spectacular developments.

It can indeed be maintained that the idea of 'solar architecture' is redundant — throughout the world there are no animal habitats which neglect those thermal parameters directly related to local insolation levels. A corollary of this proposition is the use of response patterns analogous to those of the animal kingdom. Nidification forms can indeed be a function of microclimate, in the sense meant here.

1.1.2 Description

Traditional solar practice can be reduced schematically to four elements.

(i) *Insulation*

  (a) *Radiation reflection:* albedo, surface conditions, colour, surface properties.

  (b) *Distribution of heat flow:* for example double partitioning, air circulation, external heat rejection.

2

(ii) *Exothermic reactions to solar radiation:* elevation of the temperature of a body exposed to solar radiation.

(iii) *Endothermic reactions to solar radiation:*

    (a) evaporation,

    (b) nocturnal re-radiation of solar heat.

(iv) *Storage:* required principally to compensate for the effects of nocturnal heat loss (iiib).

1.1.3 Examples in Nature

The most complex applications appear in termite nests, and it would appear that all the most desirable techniques have been brought into use.

The performances of these examples are very exact; the activity of a termitary is directly related to its temperature, as the external environment is much more hostile for termites than for humans.

Termitaries do not seem to use other energy sources to control humidity, ventilation, or temperature.

Particular examples

The Australian termite *Hamitermes Meridionalis* builds a vertical nest of laminar form (exceeding five metres in height) in which the major axis of the horizontal section is invariably in a north—south orientation, thus offering a maximum surface area to the low sun and a reduced surface to the strong sun.

Cape Colony termites have a black nest, whereas the surrounding ground is light-coloured. This colour is well-maintained and when nests are abandonded tends to fade under the effects of weathering.

The cellular structure of the termitary possesses insulation properties. These properties allow the inhabited areas of the nest to remain at $30°$ C whether the outside temperature is at $70°$ in sunlight or $10°$ at night.

In arid regions termites are capable of going to depths of over 35 metres in search of water in order to humidify the nest to such an extent that water can be squeezed out in the hand.

In the nests of a different termite (*Macroterme Natalensis*) the outer walls are more than 50 centimetres thick and provide *thermal inertia.* An example of the dependence of insect societies on insolation level is provided by European ants. At the end of winter when the nest is in a state of lethargy, some ants that are less sensitive to low temperatures and have remained awake all winter occasionally leave their hibernation chambers. As soon as the weather gets warmer and the sun shines these ants warm up again and when they re-enter the nest transfer part of this heat. Thus they contribute to the raising of the internal temperature and induce a chain reaction which is rendered possible by the unequal reactions of the insects when confronted by the cold.

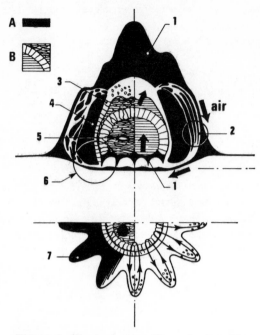

Figure 1. Thermal operation of a termitary *(Macroterma Natalensis)*.

1. First wall.
2. Gas and heat exchange zone.
3. Ventilation.
4. Insulation.
5. Second wall (royal chamber).
6. Zone for control of the heat flux by modification of the airflow in the galleries.
7. Gas and heat exchange zone.

A. Solid material — large thermal inertia.
B. Cellular material — to provide insulation.

Warm air is drawn by convection through the nest from its base to a high chamber (1). From here the air returns to the base *via* ducts (2) which pass through the outer regions of the wall where it undergoes modification of temperature, humidity, and gaseous composition

1.1.4 Solar engineering in the termitary

Even though we have not found a treatment of this question in the specialist literature, the pattern of solar engineering in the termitary (Figure 1) is quite clear and produces the mechanical behaviour (convection, etc.) which is given below.

PHASE I. Convection

(1) Cooling of the outer zones can be obtained

(a) by *evaporation,* if they are warm,
(b) by *selection,* the zones in the shadow can be used separately as exchangers.
(c) by *nocturnal* use of the external surfaces. These become cold, and the thermal flux through the outer wall allows the inner partition walls of the nest to radiate.

(2) Cool air is stored in the bottom chamber formed by the space left between the soil and the base of the nest. The nest forms a 'table' connected to the soil by conical studs; nocturnal flow is therefore possible as well as ventilation control.

PHASE II. Conduction in the mass

The bulky outer wall acts as a thermal store. Its thickness (up to 60 centimetres) permits an inversion between internal and external conditions.

The nest rests on a pedestal which completes both this outer wall and the insulation from the soil.

A cross-section of the nest reveals differing honeycomb modules in the outer layers which suggest concentric thermal walls. The texture of the central material is fairly homogeneous and contains the very massive enclosed royal chamber (several centimetres thick, with a central horizontal access which permits the passage of only one worker ant).

1.2 'TRADITIONAL' SOLAR ARCHITECTURE

It is not relevant here to describe the development of the particular design solutions adopted and we will restrict ourselves to describing the thermal behaviour of the dwellings which we have selected primarily for their illustrative value.

Before describing more detailed applications, we will discuss briefly the general use of solar energy — building orientation, exposure, etc.

In regions where walls with high thermal inertia are used, it is usually found that the traditional wall thickness is closely related to the appropriate thickness for countering the nocturnal inversion of temperature (noted by C.S.T.B. among others). The preponderance of dark colours in cold regions and bright colours in regions with strong insolation is, without doubt, significant (for example, in Canada southern walls are often painted in dark colours); also significant are the heavily-built walls in the South, insulation in the North, etc.

The Islamic world provides us with the most interesting examples for several reasons:

(a) accessibility of working examples in the Arab world (site visits),
(b) geographical position (a high-insolation area),
(c) history: the examples used are the outcome of local adaptation produced by several thousand years of technological blending,
(d) this adaptation most probably occurred during a period of 'scientific' development in the privileged areas of Islam.

Islam has developed its most elaborate thermal systems through the use of an inward facing building plan. This arrangement would seem to have originated in Ptolemaic Egypt. It spread into Greece and Asia, and at the end of the 4th Century B.C. was carried by the Greeks to Tunisia, and then to all of North Africa and the Mediterranean basin.

North African houses at the beginning of the 19th Century were based on a central courtyard surrounded by an *oecus* (from the Greek *oikos*) which formed the main room. This was the model and the Byzantine adaptation that Islam had adopted, developed and spread, even to black Africa.

1.3 SOLAR-THERMAL PERFORMANCE

The Arabic house is composed of a thermal shell which is exposed to the sun and of heavy construction (thick walls, flooring tiles, clay, etc.). This shell provides both insulation and thermal stability. Generally it has few openings to the outside. Often it has an indirect or zigzag entrance or one which is screened, and it may also be brightly coloured.

1.3.1 General external shape

Detached

Its general shape tends to be relatively tall and slim so as to reduce the roof-to-volume ratio. This was desirable because of the difficulties in supporting spans of over three metres and of ensuring watertight roofing. It also reduces the surface area of roof exposed to the sun at its zenith, thus protecting it, and it facilitates the quest for a deep courtyard.

Attached

Buildings can be united to form a block with their neighbours, offering a reduction in insolation on the roof and a compliance with the rules of town-planning.

Plan

The rooms are arranged against the inner surface of the external wall. They open on to the central courtyard which is partially or completely covered either by a roof or by a means of circulation such as a gallery, loggia, balcony, passageway, colonnade, peristyle, etc.

1.3.2 Operation

The outer wall has sufficient inertia to ensure temperature inversion in the system, the internal face being cool by day and warm by night. The radiant

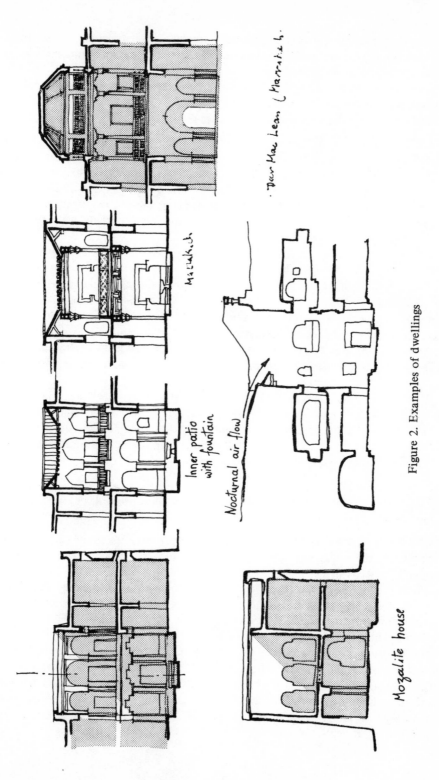

Dar Mac Lean (Marrakish)

Matbakich.

Inner patio with fountain

Nocturnal air flow

Mozalite house

Figure 2. Examples of dwellings

properties of the internal face of this wall have different uses: nocturnal heating in the cool sunny seasons, and the production of an updraught and ventilation in the hot season through a funnel sited between the wall and the junction with the flooring. However, it is through a study of the function of the courtyard that the thermal logic of these houses is most clearly seen.

The courtyard's function as a heat exchanger has two aspects:

(1) *Ventilation by the 'chimney' effect.* The upper part, exposed to the sun, is heated and induces an updraught, drawing in air from the cool lower areas (cellars, cisterns, shadow zones, garden). The heat exchanger formed by this process can operate with only a small difference in temperature if the flow is sufficient and its efficiency can be improved if a source of evaporative cooling is present (fountain, stream, pool, spray, vegetation, etc.). These systems function equally well regardless of prevailing winds, breezes, etc.

(2) *Air-conditioning by the courtyard through the thermal deficit caused by radiation losses to the night sky.* The heavy components (ground, walls, water, etc.) at the bottom of the courtyard warm up only slightly during the day and radiate to the clear sky at night, especially if they have a high emissivity (polished surfaces, marble, etc.). They can thus maintain a stable heat deficit relative to external surroundings which are less emissive and subject to insolation.

Because of its greater density the cold air mass will stay in a space having only a high opening, so that different conditions are established in such spaces open to the sky.

This cold air principle is used by all subterranean architectures designed around a 'light well' (Egypt, Siwa, Tunisia, Matmata, Spain, Benimamet (Valencia), China (Shansi), Israel (archeological excavations)). In these designs the cool air which collects in the well was produced through contact with the surface of the soil, which had been cooled by radiation to the night sky.

1.3.3. Some possibilities showing the adaptation of these principles to specific cases

The central space can either be *exposed* to or *protected* from the sun by higher elevations to the south, parapets, peripheral canopies, lattice-work plants, or even *covered* by a stone slab or a roof, possibly raised.

These sunshields protect the internal masonry as well as the space. Balconies, galleries, loggias, awnings, canopies, protrusions, ledges, claustra, trellices, and parapets all reduce the heat storage capacity of the heavy masonry.

Water is often incorporated in dry region designs, either in the central courtyard or on the principal airflow circuit. It can be *stationary,* recalling the *impluvium* of ancient Rome, when it then forms a useful heat exchanger if the water mass is sufficient and if the walls of the pool have a high emissivity, enhancing the nocturnal cooling. It can *flow* in channels, or *trickle* in a thin film over a solid mass of masonry, either over sloping surfaces as at Sil Sibil, or in *fountains* or *sprays,* for example to water vegetation.

2
General Principles

2.1 THE LAWS OF SOLAR RADIATION

2.1.1 The sun

The sun is a gaseous sphere of helium and hydrogen with small amounts of carbon and other elements.

Its average density is 1,400 kg m^{-3} (5,500 kg m^{-3} for the earth) but it reaches 76,000 k m^{-3} in the centre. Its radius is 700,000 km (the earth's radius is 6,400 km).

Its mass represents 99.85% of the total mass of the solar system and is 33,000 times that of earth.

Its age has been estimated as about six thousand million years and it will probably last for a further similar period.

The sun is a rotating mass, whose equatorial region rotates in 24 days while the regions nearest to the poles rotate in 30 days.

As far as man is concerned, the sun can be considered an inexhaustible energy source.

2.1.2 Solar radiation

According to Albert Einstein, energy and mass are equivalent and are related by
$$E = mc^2 \tag{1}$$

where E = energy
m = mass
c = speed of light (300,000 km s^{-1})

In the centre of the sun, nuclear reactions are produced: hydrogen is turned into helium releasing 4 million tonnes of mass energy per second.
4 hydrogen → 1 helium + conversion into energy of 1/141 of the mass involved in the reaction.

This corresponds to 2 cal cm^{-2} min^{-1}. In order to avoid collapsing inwardly under

9

its own gravitational pressure the sun radiates this energy and an infinitesimal part is collected by the illuminated surface of the earth.

Surface temperature of the sun

The sun radiates energy according to the Maxwell–Boltzmann law of black body radiation:

$$E = \sigma T^4 \tag{2}$$
$$\sigma = 4.9 \times 10^{-8} \text{ kcal h}^{-1} \text{ K}^{-4}$$
T = absolute temperature in degrees Kelvin ($= 273.15 + t°C$)

As E is known for the hydrogen–helium reaction one can calculate T:

$$T = 5{,}762 \text{ K.}$$

2.1.3 The solar constant

The average distance from the earth to the sun is $d = 149$ million kilometres.

The orbit of the earth around the sun is a slightly eccentric ellipse.

At the summer solstice (21st June), the earth is at a maximum distance from the sun of $d_{max} = 1.017 \times d$.

At the winter solstice (22nd December) by contrast, it is at its minimum distance of $d_{min} = 0.983 \times d$.

The sun's disc subtends an angle of $32'$ of arc (about ½ degree) at the earth's surface.

The solar constant, that is, the direct radiation I_0 received by unit surface area normal to the solar radiation at the upper limit of the earth's atmosphere, is 1,353 W m^{-2}. This figure is the result of satellite and rocket measurements.

The variation of the earth–sun distance introduces a variation in the value of I_0.

The solar constant I_0 represents the total radiation in the solar spectrum which has the form shown in Figure 5.

Thermal radiation is defined as the radiant energy emitted by a body due to its temperature.

The emission of thermal radiation is governed by the temperature of the emitting body. (The sun: 5,469 °C; terrestrial bodies: 20 °C or more). The spectral range lies between 0.1 and 100 μ ($1 \mu = 10^{-6}$ m).

The solar radiation emitted at 5,469 °C lies between 0.2 and 5 μ

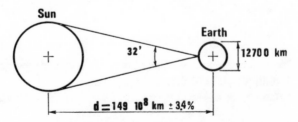

Figure 3. The distance between earth and sun

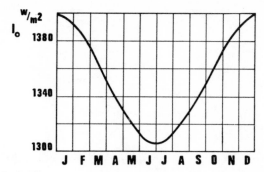

Figure 4. Variation in value of the solar constant I_0 throughout the year

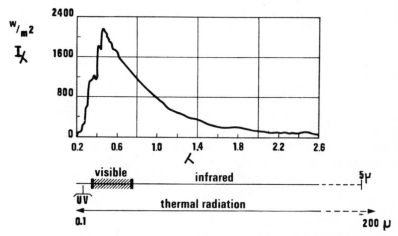

Figure 5. Solar radiation spectrum, as a function of wavelength λ expressed in μm $(1\,\mu m = 10^{-6}\,m)$. Area under the curve is $I_0 = 1.353$ W m^{-2}

2.2 ATMOSPHERIC PHENOMENA AND THEIR INFLUENCE ON RECEIVED SOLAR ENERGY

The atmospheric mass traversed by the solar radiation is equivalent to that of a uniform atmosphere 8 km thick whose pressure and temperature are constant and equal to the sea level value.

The solar radiation scattered and absorbed by the gas molecules and atmospheric particles undergoes an attenuation whose size depends upon the number of molecules and particles encountered by each ray.

As the optical path of the solar radiation varies depending on the angular height of the sun above the horizon, one can introduce the idea of atmospheric mass traversed by the radiation, denoted by m.

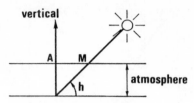

Figure 6. Atmospheric air mass

The unit of air mass $m = 1$ corresponds to OA, the vertical through the average atmosphere with a pressure at the ground of 1,000 millibars (sea level). From this, at a given place, if the atmospheric pressure differs from 1,000 millibars, the atmospheric air mass will be: $OA = m = P/1,000$ where P is the atmospheric pressure at the location under consideration.

The attenuation of radiation is a minimum during a passage from the zenith. If h is the angular height of the sun, the radiation path is

$$OM = \frac{OA}{\sin h} = \frac{m}{\sin h} = \frac{P}{1,000 \sin h}$$

If, for example, $h = 30°$ OM will be equal to 2 and the number of particles and molecules will be double that corresponding to a vertical path through the atmosphere.

The absorption by the atmosphere gases (ozone, carbon dioxide, water vapour) can be represented by a thickness of liquid water showing the same absorption characteristics, represented by the coefficient W. One can take, for example, for our latitudes the following values:

$W = 1$ cm in winter,
$W = 2$ cm in summer.

Molecular diffusion can interfere with the optical path characterized by m.

Diffusion by aerosols introduces Angström's turbidity factor β, which can take the following values:

$\beta = 0.05$ under blue sky,
$\beta = 0.10$ under an average sky,
$\beta = 0.15$ to 0.20 in areas with urban pollution.

A knowledge of the coefficients (W and β) allows us to choose standardized conditions.

One can thus obtain the values of direct radiation I received at the ground:

Angular height h	5°	10°	20°	30°	40°	60°	90°
I when $W = 2$ cm and $\beta = 0.08$	132	355	610	748	860	920	950

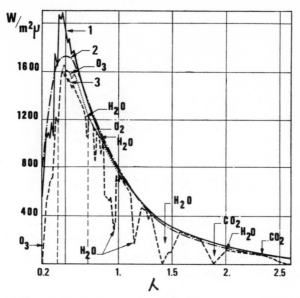

Figure 7. Radiation spectra outside the atmosphere and at sea level.
1. Solar radiation outside the atmosphere.
2. Radiation from a black body at 5,762 K.
3. Solar radiation at sea level; absorption bands by the molecules O_3 (ozone), H_2O (water vapour), CO_2 (carbon dioxide)

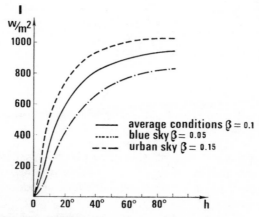

Figure 8. Variation of direct radiation I with the angular height of the sun h for different atmosheric conditions

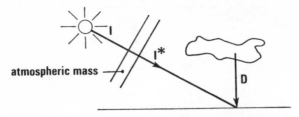

Figure 9. Direct radiation I^* received at the ground

2.3 MEASUREMENT OF SOLAR RADIATION AT GROUND LEVEL

2.3.1 Clear sky radiation at ground level

The radiation received is the total radiation G_{OH} where O indicates the clear sky and H the horizontal. It consists of two terms, the direct radiation I_0^* (I_0 corrected for the atmospheric factors), and the diffuse radiation D_0 from the whole sky.

$$G_{OH} = I_0^* \sin h + D_{OH} = S_{OH} + D_{OH}.$$

Total radiation is measured by a pyranometer.

A knowledge of G_{OH} is insufficient for calculating the solar gains on any surface as the diffuse component of the radiation must be known separately. For example, one cannot apply to G_{OH} those operations which allow the transformation from a horizontal plane to a vertical plane. The diffuse radiation behaves differently; a vertical surface *sees* only one half of the sky and also intercepts that part of the total radiation reflected by the ground near the surface of interest (albedo).

The albedo depends on the character of the ground and represents that fraction of total radiation reflected towards the surfaces under consideration.

Meteorological measurements show that D_{OH} for the winter period is equal to 0.30 G_{OH}; it can reach 40% of G_{OH} in December in the Paris region. Following from this, one can determine the value of G_{OH} by the difference, whence $I_0^* = S_{OH}/\sin h$.

2.3.2 Clear sky radiation on an inclined surface which is at a general orientation with respect to due south

For the direct component, one can start from a knowledge of S_{OH} (see Section 2.5 and Fig. 12) and multiply $S_{OH}/\sin h$ by the projection factor $\cos u$ (for examples see Sections 2.5.6 and 4.8) where u is the angle between the normal to the plate and the sun's rays.

For the diffuse component, the inclination i of the surface determines the fraction of the sky and the portion of the total radiation G_{OH} reflected by the ground that are seen by the surface.

$$D_0 = D_{OH}a + \alpha b G_{OH} \tag{7}$$

where

$$a = \frac{1 + \cos i}{2}, \quad b = \frac{1 - \cos i}{2}, \quad \alpha = \text{albedo coefficient}$$

$\alpha = 0.18$ to 0.23 for grass
$\alpha = 0.55$ for concrete
$\alpha = 0.18$ for macadam
$\alpha = 0.80$ to 0.90 for fresh snow.

For a vertical surface, taking $\alpha = 0.3$

$$D_{OV} = 0.5(D_{OH} + 0.3 G_{OH}) \quad (V \text{ implies vertical}). \tag{8}$$

2.3.3 Average Conditions

The relation

$$G_H = I^* \sin h + D_H = S_H + D_H \tag{9}$$

always holds (with or without the subscript O).

In order to determine D_H one must know the insolation fraction σ where

$$\sigma = \frac{\text{number of hours of sunshine}}{\text{maximum number of possible hours}} \quad (\text{see Section 2.6.7}),$$

and one has the relation

$$D_H = G_H (1 - 0.25\sigma - 0.60 \sqrt{\sigma}). \tag{10}$$

To project onto a general surface we proceed in the same way as before. For numerical examples see Sections 2.5.6 and 4.8, and to determine G_H one can use the relation

$$G_H = G_{OH} (0.33 + 0.7\sigma). \tag{11}$$

2.3.4 Approximation of the radiation curve by a sinewave

The maximum intensity of received radiation occurs at local noon. The radiation curve can be approximated by a sinewave whose area equals total energy received on a given day.

$$\underset{\substack{\downarrow \\ \text{for} \\ \text{1 day}}}{Q} = \frac{2}{\pi} \times \underset{\substack{\downarrow \\ \text{maximum} \\ \text{value at} \\ \text{local noon}}}{G} \times \underset{\substack{\downarrow \\ \text{hours in} \\ \text{the day}}}{\Delta T} \tag{12}$$

(for an example see Sections 2.5.6 and 4.8).

2.3.5 Evaluation of radiation in the absence of experimental data

In the absence of experimental data for a site, the direct insolation can be estimated using the equation

$$I^* = rI_0 p^m \tag{13}$$

where

I_0 = the solar constant = 1,354 W m^{-2}

p = the atmospheric transmission factor = 0.71

(for towns p = 0.6 and for mountain regions p = 0.85).

$$m = \frac{1}{\sin h} \times \frac{P}{1,000}$$

where P depends on the altitude of the site

r = the percentage correction of the Earth—Sun distance
  + 3.4% at the winter solstice
  − 3.4% at the summer solstice.

2.4 MEASUREMENT APPARATUS

The information necessary to determine the solar energy potential of a given site includes: the hours of sunshine, the direct and diffuse radiation levels, and the total radiation.

2.4.1 Sunshine indicators

To determine the hours of sunshine, sunshine indicators (or heliographs) record the periods during the day for which the intensity of direct radiation exceeds a certain level. The sum total of these periods represents the daily sunshine hours. The time interval between sunrise and sunset defines the maximum number of sunshine hours at the selected site on that day.

The insolation fraction is the ratio of the measured daily insolation to the theoretical maximum. Several types of heliograph give this daily insolation. The

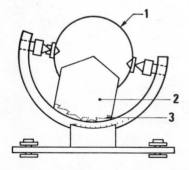

Figure 10. Campbell—Stokes Sunshine Indicator. (1) The glass sphere focuses the radiation at a point on the special paper (2) producing a burnt trace. As it moves the burning point traces a curve (3), whose length is proportional to the duration of the sunshine

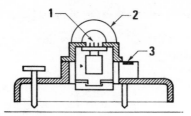

Figure 11. Pyranometer
(1) Thermopile.
(2) Double glass hemisphere (to suppress convection currents).
(3) Level

detector focuses the direct radiation by means of a focusing lens which is generally a sphere of solid glass. This will record irrespective of the orientation of the sun to the instrument during the day. The record is made either on plain paper which is burnt black when the direct radiation intensity exceeds the chosen level (the Campbell—Stokes sunshine recorder), or on a photographic paper (the Jordan sunshine recorder). These two instruments are not provided with a timing trace since the timescale can be determined from the length of the recordings themselves.

2.4.2 Pyroheliometers

Pyroheliometers quantify the direct radiation. Thus, the silver disc pyroheliometer allows the measurement of the intensity of direct radiation to be determined from successive thermometric readings with the aperture of the apparatus alternately opened and closed. Certainly such a device should be furnished on any automatic tracking system.

2.4.3 Pyranometers

Pyranometers, which are usually associated with continuous recording, as in a thermopile for example (see Figure 8), give the total radiation (direct and diffuse).

In order to eliminate the direct radiation, some pyranometers, like the one used at the C.N.R.S. solar furnace at Odeillo, are fitted with a mask which eliminates the direct component. While these instruments are, by and large, sufficient for solar measurements aimed at domestic or environmental applications they are unsuitable for more specific studies because they are non-selective in regard to wavelength. Other instruments, such as spectrobolometers, will give the same parameters for the different spectral bands (ultraviolet, visible, near infrared).

2.5 THE HEIGHT OF THE SUN IN THE SKY AT A GIVEN PLACE: EXPRESSIONS FOR RADIATION RECEIVED ON A GENERAL SURFACE

The height of the sun is a fundamental parameter for two reasons:
(a) This height appears explicitly in all calculations for energy gains on surfaces, as we will see later.

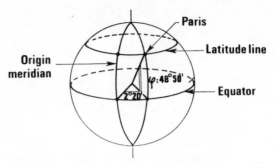

Figure 12. Plotting a point on the earth's surface

(b) The planning of a site requires a knowledge of the shadows cast by existing buildings and vegetation or due to the relief, as in mountain areas, which could screen the collectors after the completion of the building. It is therefore very important on this point to compile tables or charts of the height of the sun for each month.

2.5.1 Latitude, declination, and hour angle

A point on the surface of the earth is represented by two angular coordinates, *latitude and longitude.*

The origin of longitude is Greenwich, England at $0°$. Places to the east are described with the positive sign (e.g. Strasbourg $+7°38'$ East). Those to the west have a negative sign (e.g. La Rochelle $-1°09'$ West). Alençon is practically at longitude $0°$.

The signs $+$ or $-$ occur in the determination of the true solar time.

The arc of the circle joining the North Pole, Greenwich, and the South Pole is the origin meridian. There are 23 other meridans, each separated by $15°$ ($24 \times 15° = 360°$), producing the 24 time zones.

The latitude ϕ allows us to refer the angular distance of any point to the equator. It varies from $0°$ on the equator to $+90°$ at the North Pole and $-90°$ at the South Pole.

Solar declination δ

The earth rotates around the sun in an elliptical orbit lying in a plane (the plane of the ecliptic); the normal to this plane and the axis of rotation of the earth make a variable angle δ that is called the solar declination and which has the following values:

$\delta = +23°27'$ at the summer solstice (22nd June),
$\delta = -23°27'$ at the winter solstice (22nd December),
$\delta = 0°$ at the spring and autumn equinoxes (21st March and 23rd September).

At the equinoxes day and night are of equal duration.

Between these four special points, δ takes all the intermediary values which can be calculated by the relation

$$\sin \delta = 0.4 \sin t \tag{14}$$

where t corresponds to the number of days (N) since the spring equinox multiplied by the correction factor 360/365.

$$t = N \times \frac{360}{365}.$$

Average monthly declination of the sun

	January	February	March	April	May	June
Monthly Average	−20.8	−12.7	−1.9	+9.9	+18.9	+23.1

	July	August	September	October	November	December
Monthly Average	+21.3	+13.7	+3.0	−8.8	−18.4	−23.0

Hour Angle HA of the sun

A day corresponds to the time separating two successive journeys of the sun across the same meridian.

The hour angle HA is therefore determined by the regular rotation of the earth around its axis; consequently it is directly linked to real solar time.

The hour angle takes the following values:

HA = $0°$, when the sun crosses the meridian plane of a location; it defines local noon.

The relation between time and angle:

1 hour of time = $15°$ of hour angle
1 minute of time = $15'$ of arc
1 second of time = $15''$ of arc
HA = $90°$ at 18.00 h (local solar time) and $−90°$ at 06.00 h.

Hour angles are calculated with the sign + from local noon until sunset and with the sign − from sunrise until noon.

2.5.2 Clock times and local solar time

For reasons which are readily understood, in most countries clocks show the same time if possible. The following equation converts clock time to local solar time (LST).

LST = local clock time − N (number of time zone, e.g. No. 1 = 1 hour, No. 2 = 2 hours) + correction for longitude in mins. + time correction in mins.

(a) Correction for longitude

This occurs when one considers places at different longitudes within the same zone. For example:

Brest 4°28′ West longitude

Strasbourg 7°38′ East longitude

These towns are separated by about 12°6′ which, when converted, corresponds to 54 minutes of time difference (1° = 4 mins.).

This is the time difference separating local noon for the two towns.

(b) Time correction

In order to regulate our clocks, the real sun is replaced by an imaginary sun which in its apparent movement with respect to the earth remains in a plane. In fact, the plane of a solar trajectory follows a variable angle δ, the solar declination which we have already defined.

The deviation between the imaginary sun and the real sun is determined by the time equation whose values fluctuate between ±15 minutes. Longitudinal or nautical atlases supply the necessary information.

Changing notation, local solar time can be written:

$$T_{CO} = T_{CN} + N$$

where N is the number of the time zone, positive from No. 1 to 12, negative thereafter.

2.5.3 Examples

If it is 14.00 h in Paris, civil time, on 15th April, what is the local solar time?

(a) Although Paris is in the same time zone as Greenwich, since 1940 France has advanced her clocks by 1 hour in alignment with her continential neighbours, Therefore, $N = 1$ (1 hour of correction).

(b) Correction for longitude: Paris is 2° East longitude, 2° = 8 minutes of time.

(c) Correction for the time equation: on 15th April the correction is 0 minutes.

Therefore, LST = 14.00 h − 1 h + 8 min + 0 min = 13.08 h.

The hour angle of the sun which determines its position in the sky is

$$HA = 1 \text{ h } 8 \text{ min of time} = 15° + 2° = 17°.$$

2.5.4 Coordinates of the sun in the sky

The position of the sun in the sky for a given place and a given time can be represented in either of two systems of coordinates.

The more simple is the system of horizontal coordinates, made up of the horizontal plane and the normal to this plane. The second is the system of equatorial coordinates which we will not consider further. The sun's movement is represented by taking the site under consideration as the origin and the centre of a sphere (the celestial sphere). The equatorial plane and the axis of this sphere are the

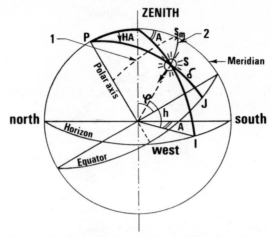

Figure 13. Coordinates of the sun in space.
ϕ = latitude, HA = hour angle, δ = declination,
A = azimuth which allows the track of the sun
to be plotted on the horizontal plane (origin
HA = 0°), $A = 0°$, West $A = 90°$, North
$A = 180°$, East $A = -90°$.
(1) The trajectory of the sun in the morning.
(2) The trajectory of the sun in the afternoon

horizontal plane and the vertical at the site. We can now use the different angles as defined above and in Figure 13. In particular for the spherical triangle P-zenith-S there are the following relations,

Height of the sun

$$\sin h = \sin \phi \sin \delta + \cos \phi \cos \delta \cos \text{HA} \tag{15}$$

Track of the sun in the horizontal plane: it is necessary to calculate the azimuth as the receiving surface is not horizontal.

$$\sin a = \frac{\cos \delta \sin \text{HA}}{\cos h} \tag{16}$$

Important values of Equation 15

(a) at local noon HA = 0, therefore cos HA = 1

$$\sin h = \cos \left(\frac{\pi}{2} - h \right) = \cos(\phi - \delta) \tag{17}$$

$$h_{max} = \frac{\pi}{2} - \phi + \delta$$

(b) sunrise and sunset:

$$h = 0, \text{ therefore } \sin h = 0$$

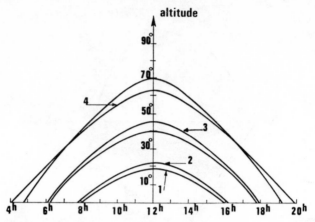

Figure 14. Curves of corresponding altitudes for different months at Paris and Carpentras.
(1) 21st December at Paris.
(2) 21st December at Carpentras.
(3) 21st March at Carpentras.
(4) 21st June at Paris

$$\cos AH_0 = -\frac{\sin \phi \sin \delta}{\cos \phi \cos \delta} = -\text{tg } \phi \text{ tg } \delta \tag{18}$$

$$AH_0 = \text{arc } \cos(-\text{tg } \phi \text{ tg } \delta)$$

h and HA can be determined with the aid of trigonometric tables or with a calculator having the functions sine and cosine.

As the sun only rises and sets exactly at the East and West on two days of the year, at the equinoxes, it is necessary for all other periods to have curves of the deviation from these two important points.

To do that, a stereographic projection is used which consists of projecting the

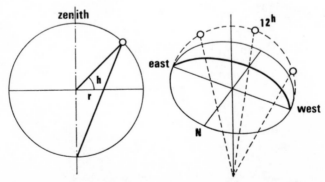

Figure 15. Stereographic projection applied to a track of the sun for an equinoxial day

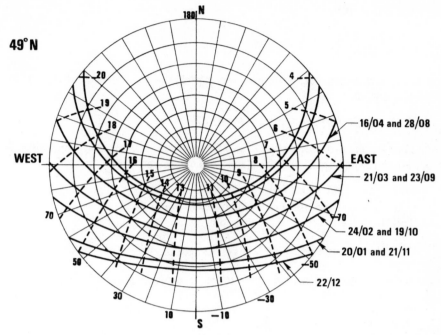

Figure 16. Graph of sunshine for the latitude 49 °N (Paris)

sun's trajectory onto the equatorial plane of the celestial sphere (the local horizontal).

On the horizontal plane we plot for each hour of the day the intersection with the plane of the straight line joining the sun to the other end of the diameter through the zenith (see Figure 1).

Example: on 21st March at 08.00 h the azimuth of the sun is $-66°$ and the height is $19°$. (See Figure 16. For each latitude there is a similar graph.)
Applications: Study of the geography of the site,
Determination of the shadows of other buildings, vegetation, etc. (See Section 2.7)

2.5.5 Radiation received on a general plane

It has been shown that the total radiation consists of two components: the direct (I^*) and the diffuse (D).

i is the angle of inclination of the normal of the plane with the local vertical. γ is the angle of incidence of the direct radiation with the normal to the plane.

The useful component is given by the projection onto the normal,

$$S = I^* \cos \gamma \tag{19}$$

For a horizontal plane,

$$\gamma = 90° - h$$
$$S = I^* \sin h \tag{20}$$

For a vertical plane, facing south, the normal is the southerly direction and projecting onto it,

$$S(90° \text{ South}) = I^* \cos h \cos a = I^* \cos u \tag{21}$$

For a plane inclined to the horizontal and facing south, it is necessary to project the insolation onto the normal and there are two components for I^*:

  onto the vertical, $I^* \sin h$, and
  onto the horizontal, southerly direction, $I^* \cos h \cos a$
  Therefore, for the normal to the plane

$$S = I^* (\cos h \cos a \sin i + \sin h \cos i) = I^* \cos u \tag{22}$$

For a vertical plane in a general orientation, by projecting onto the normal to the plane,

$$S(90° \; \alpha) = I^* \cos h \cos(a - \alpha) \tag{23}$$

$\alpha = -45°$ for south-east façade
$\alpha = 90°$ for a west façade
$\alpha = -90°$ for an east façade

For a plane of general orientation and inclination,

$$S(i, \alpha) = I^* (\cos h \sin i \cos(a - \alpha) + \sin h \cos i) \tag{24}$$

$$S(i, \alpha) = I^* \cos u$$

In these expressions I^* is determined by the known formulae for the laws of solar radiation.
  G can be calculated from

$$G = S + \text{Diffuse}$$

The formulae given above show that for a period of any length it is advantageous to make the calculations on a programmable calculator.

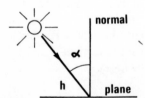

Figure 17. The relationship of the angles and the normal to the horizontal plane

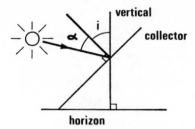

Figure 18. Collector inclined at an angle *i* to the horizontal

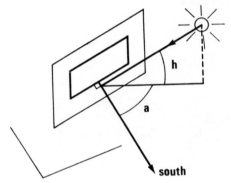

Figure 19. Vertical plane oriented towards the south

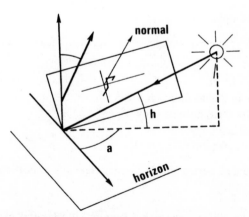

Figure 20. Inclined plane oriented towards the south

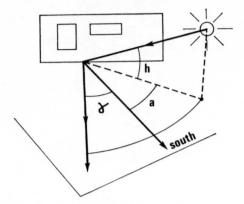

Figure 21. Vertical plane of general orientation

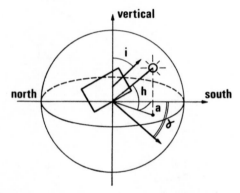

Figure 22. Plane of general orientation and inclination

2.5.6 Numerical example

To calculate the received energy on a vertical façade facing south on 15th December at Carpentras:

Basic data: latitude$-44°$N
altitude of sun at noon$-h = 23°$
azimuth of sun at noon$-a = 0°$
length of day$-\Delta T = 8.5$ h
insolation fraction$-\sigma = 0.58$.

Total radiation recorded during the day under clear sky conditions:

$$G_{OH} = 369 \text{ W m}^{-2}.$$

Calculation of the diffuse component on the horizontal plane: this represents 30% of total radiation for one day under clear sky conditions.

$D_{OH} = 369 \times 0.30 = 111$ W m^{-2}

and thus the direct component on the horizontal plane is given by Equation 6 as

$S_{OH} = I_0 \sin h = G_{OH} - D_{OH} = 258$ W m^{-2}.

To calculate the direct component on the vertical plane Equation 21 is written

$$S_{OV} = S_{OH} \times \frac{\cos u}{\cos h}.$$

Trigonometric tables give: $\cos h = 0.9205$

$\qquad\qquad\qquad\qquad\quad \sin h = 0.390$

$\qquad\qquad\qquad\qquad\quad \cos a = 1$

whence

$$S_{OV} = 258 \times \frac{0.9205}{0.390} = 608 \text{ W m}^{-2}.$$

To calculate the diffuse component on the vertical plane we have

$D_{OH} = 0.5(D_{OH} + \alpha G_{OH})$.

If the albedo α is 0.3

$D_{OH} = 111$ W m^{-2},

therefore, the total radiation on the vertical plane per day under clear sky conditions is

$G_{OV} = 608 + 111 = 719$ W m^{-2}.

To obtain the total radiation under average conditions one can use

$G_V = G_{OV} (0.33 + 0.7\sigma)$ or $G_V = G_{OV}\sqrt{\sigma}$

which gives

$G_V = 719(0.33 + 0.7 \times 0.58) = 529$ W m^{-2}.

Thus the energy received during the day is (Equation 12)

$$Q = \frac{2}{\pi} \times G_V \times \Delta T = 2{,}864 \text{ Wh m}^{-2}.$$

2.6 METEOROLOGICAL DATA FOR A GIVEN SITE

2.6.1 Climatic scale

Climate is determined by the state of the atmosphere for the latitude of the place being considered and of its substratum (continent, oceans). From meteorological data one can specify the size of each parameter by studying the averages calculated over long periods (10 years) and in different places.

An atmospheric state is characterized by the following data or parameters:

air temperature
windspeed
humidity of the air
insolation (hours of sunshine per day)
intensity of solar radiation (received on a horizontal surface)
precipitation.

All of these parameters are recorded daily either continuously or at fixed hourly intervals. They provide a mass of data which can be used for statistical analysis. The results are often published in the form of maps and tables.

Sometimes algebraic expressions can be used to smooth the curves produced by the statistical studies and are directly calculable using a pocket calculator.

2.6.2 Climate and environment of buildings

Climatology distinguishes three scales: the *macro-* or *regional scale* (75 to 100 km); meteorological parameters are measured and recorded in stations of the network. The published data and the climatic classification represent the monthly and annual figures for the climatological compontents.

local scale (town, valley); the meteorogical data to be used for a study must be extrapolated from the data from the nearest meteorological stations and completed by observations or investigations on models.

micro-scale; on this minute scale, the influence of the buildings themselves is predominant. The interactions between the immediate external environment and the physical environment of the buildings are directly influenced by the structure itself.

Example: the effects of wind around tower blocks. It has proved necessary to allow for 'wind-traps' around the buildings as the gap between two towers can behave as a funnel and accelerate the air streams.

In the case of a particular site plan, the data published by a meteorological station is insufficient. It is necessary to provide a further series of observations to determine the prevailing wind, the effects of other buildings and the role of vegetation (wind breaks or sun shields).

2.6.3 The French meteorological network

Meteorological data is measured at a certain number of stations distributed throughout the whole country. However, all are not of practical interest for solar applications, which depend to a large extent on the particular utilization programme for the solar energy. In our latitudes heating of dwellings and the production of domestic hot water would appear to be the most advantageous applications, whereas for lower latitudes (e.g. Athens) the problem is rather to ensure thermal comfort in summer by air-conditioning by means of solar energy.

In order to utilize solar energy two factors are indispensable: radiation intensity and duration of insolation.

Equally important are heat losses from the building and prevailing wind conditions (site planning).

The hygrometric data is not directly necessary except for air-conditioning projects in the south of France.

2.6.4 Meteorological stations in France

At the present time there are 100 stations distributed in 14 groups (see map), where the following data is collected: air temperature, relative humidity, water vapour pressure, cloud cover, and insolation.

	Measurement centres for total radiation														
Stations	62	63	64	65	66	67	68	69	70	71	72	73	74	75	76
Agen															
Ajaccio															
Caen															
Carpentras															
La Rochelle															
Limoges															
Mâcon															
Millau															
Montpellier															
Nancy															
Nice															
Reims															
Rennes															
Saint-Quentin															
Strasbourg															
Trappes															
Tours															
Auxerre															
Dijon															

Years for which statistics are available

Thus, there is a meteorological station sited about every 75 km. The map also shows that only 19 stations record solar radiation intensity and only two (Trappes and Carpentras) measure direct and diffuse radiation.

2.6.5 Presentation of meteorological results

Solar radiation intensity and duration of insolation

(a) Solar radiation on a horizontal plane. A reading is made every hour between 05.00 h and 19.00 h (L.S.T.). Every ten days the hourly averages are calculated.

The same procedure is followed on a monthly basis and the tables give the intensity in $J \ cm^{-2}$, whence the conversion, in $kcal \ m^{-2}$

$$1 \ J \ cm^{-2} = 2.39 \ kcal \ m^{-2}.$$

Station: La Rochelle
Latitude: 46°09′ N
Longtitude: 01°09′ W
Altitude: 9.95 m
Month: January 1975

Monthly record of total

RADIATION *G* ON A
Pyranometer: KIPP

Units used: J cm^{-2}(1 J cm^{-2}

RADIATION – Hourly intervals in local solar time

Day	04 05	05 06	06 07	07 08	08 09	09 10	10 11	11 12	Morning	12 13	13 14
1	000	0	0	002	034	078	114	131	359	131	111
2	000	0	0	003	032	077	071	078	281	086	065
3	0	0	0	002	008	038	056	119	223	106	096
4	0	0	0	001	108	027	049	055	140	076	059
5	0	0	0	001	009	017	022	032	081	029	024
6	0	0	0	001	006	019	025	035	088	031	028
7	0	0	0	001	005	019	036	027	088	021	022
8	0	0	0	102	021	039	043	037	142	053	043
9	0	0	0	002	015	057	107	124	305	127	111
10	0	0	0	001	007	025	084	116	235	034	045
T$_D$	0	0	0	016	149	396	627	754	1942	724	605
11	0	0	0	002	014	027	063	064	170	065	060
12	0	0	0	004	020	051	052	076	233	055	027
13	0	0	0	002	019	064	101	114	300	124	110
14	0	0	0	001	013	030	036	032	112	026	024
15	0	0	0	001	017	014	029	044	105	064	032
16	0	0	0	000	003	020	034	058	115	075	102
17	0	0	0	002	011	017	028	021	081	015	013
18	0	0	0	002	024	038	087	074	125	141	055
19	0	0	0	002	043	071	128	148	414	137	124
20	0	0	0	006	018	038	052	038	152	023	028
T$_D$	0	0	0	024	122	390	640	671	1907	625	605
21	0	0	0	002	017	035	058	087	209	094	056
22	0	0	0	002	007	015	034	031	090	029	024
23	0	0	0	001	007	036	029	047	120	061	034
24	0	0	0	007	027	054	089	087	264	108	080
25	0	0	0	002	015	025	015	020	079	021	021
26	0	0	0	009	032	024	028	021	114	015	017
27	0	0	0	002	023	037	073	006	231	109	063
28	0	0	0	002	007	010	023	032	074	019	019
29	0	0	0	001	007	032	033	046	119	050	033
30	0	0	0	002	014	041	067	108	232	146	104
31	0	0	0	004	022	024	060	106	215	108	137
T$_D$	0	0	0	034	178	335	520	581	1748	761	588
T$_M$	0	0	0	044	509	1141	1787	2106	5597	2210	1798
M$_{av}$	0	0	0	002	016	036	058	068	170	068	058

radiation and sunshine hours

HORIZONTAL SURFACE

= 0.239 cal cm^{-2})

INSOLATION
Heliograph: Campbell Stokes

Units used: tenths of hours

14 15	15 16	16 17	17 18	18 19	19 20	Evening	Total	Morning	Evening	Total
078	032	001	0	0	0	353	712	40	41	080
040	015	002	0	0	0	208	481	35	15	050
033	004	0	0	0	0	239	462	13	19	032
056	022	001	000	000	000	024	364	00	12	00
017	007	001	0	0	0	075	159	0	0	0
021	009	002	0	0	0	041	179	0	0	0
030	018	002	0	0	0	023	181	0	0	0
022	014	001	0	0	0	133	275	01	0	001
079	034	003	0	0	0	354	659	24	40	064
043	023	001	0	0	0	177	414	16	0	256
429	175	014	0	0	0	1950	3392	129	127	256
040	030	003	0	0	0	198	363	01	02	003
016	006	002	0	0	0	106	339	15	02	017
078	026	003	0	0	0	341	641	21	37	058
017	010	001	0	0	0	078	190	00	00	0
031	006	001	0	0	0	164	269	01	00	001
048	039	002	0	0	0	266	321	00	20	020
008	006	002	0	0	0	044	125	00	00	0
043	009	003	0	0	0	151	376	10	07	017
086	042	005	0	0	0	394	308	41	37	078
015	005	003	0	0	0	074	226	00	00	0
382	179	025	0	0	0	1836	3723	90	105	195
089	025	005	0	0	0	269	478	05	20	025
028	013	002	0	0	0	096	186	0	0	0
014	006	000	0	0	0	115	235	0	0	0
063	025	005	0	0	0	281	545	03	03	006
033	024	005	0	0	0	104	183	0	0	0
018	012	002	0	0	0	065	179	0	0	0
074	022	005	0	0	0	273	504	13	19	032
012	006	002	0	0	0	058	132	0	0	0
021	017	007	0	0	0	928	247	0	0	0
047	029	005	0	0	0	331	563	04	15	019
087	052	004	0	0	0	388	604	04	35	039
486	231	042	0	0	0	2108	3856	29	92	121
1097	588	081	0	0	0	5874	11471	228	324	572
042	019	003	0	0	0	190	370	08	10	018

Figure 23. French network of solar radiation recording centres.
● Stations recording sunshine hours.
▲ Stations recording sunshine hours and solar radiation

To obtain the result in kilowatts per hour per metre squared, the following relation is used

$$1 \text{ kcal m}^{-2} = 1.16 \text{ Wh m}^{-2} = 1.16 \times 10^{-3} \text{ kWh m}^{-2}$$

The energy received on a horizontal or inclined surface can be expressed in the same form and we include in a table a chart for January.

Example: At La Rochelle in March 1971, the average energy received between 12.00 h and 13.00 L.S.T. is 174 J cm^{-2} or

$$174 \text{ J cm}^{-2} = 174 \times 2.39 \times 1.16 \times 10^{-3} \text{ kWh m}^{-2}$$

$$= 0.482 \text{ kWh m}^{-2}.$$

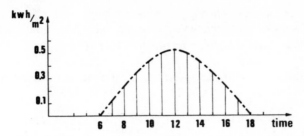

Figure 24. Energy received per day on a horizontal plane at La Rochelle, March 1971

It is possible to do the same for each hour and to plot a graph of energy received on a horizontal plane.

The total energy received will be the area under the curve, from which the daily average for March 1971 at La Rochelle was 3.5 kWh m^{-2}. The sinusoidal approximation gives 3.68 kWh m^{-2}, which is an error of just over 5%.

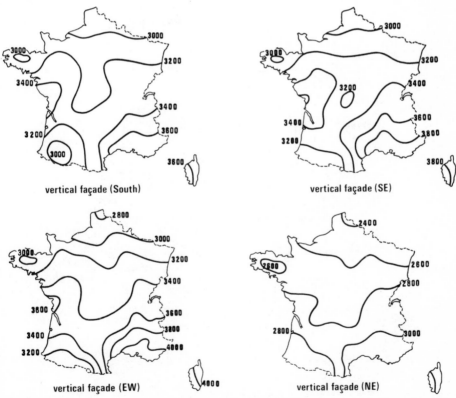

Figure 25. Energy received on vertical façades of different orientations for July (W m^{-2})

34

(b) Sunshine hours (SS) in France. Monthly and annual charts can be drawn up from measurements collected over long periods. The hours of sunshine are measured with a Campbell-Stokes recorder which has a sensitivity of approximately $100\,W\,m^{-2}$.

Example: Embrun in the Briançonnais has 2,660 hours of sunshine per year.

2.6.6 Maximum duration of sunshine (SS$_0$)

This is obviously equal to the length of the day. We have seen that at sunrise and sunset $h = 0$ and that the hour angle of sunrise and sunset is given by HA$_0$ = arc cos($-$tg ϕ tg δ).

At the equinoxes day and night are equal and of 12 hours duration. A sinusoidal variation can be adopted for SS$_0$ which must be a function of latitude (ϕ) and the number of days t since 21st March. Thus,

$$SS_0 = 12h + \phi \sin \frac{360}{365} t.$$

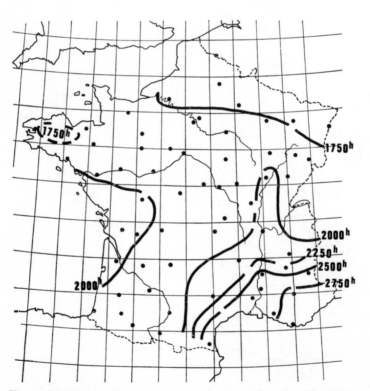

Figure 26. Average annual sunshine hours in France. The points indicate the reporting stations. (Based on a Météorologie Nationale publication)

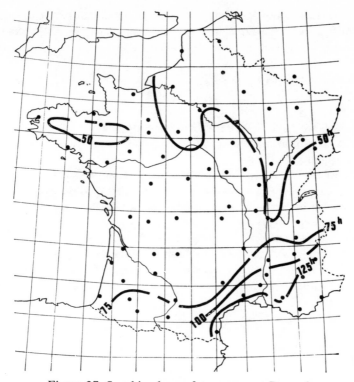

Figure 27. Sunshine hours for an average December

Latitude	ϕ	Maximum duration 21st June (summer solstice)	Minimum duration 21st December (winter soltice)
50°	4.1	16.5	7.9
48°	3.8	16.1	8.2
46°	3.6	15.8	8.4
44°	3.3	15.6	8.7
42°	3.1	15.1	8.9
40°	2.8	14.8	9.2

NOTE: The maximum number of hours of sunlight is 365 days × 12 h = 4,380 h.

2.6.7 The insolation fraction σ

This parameter gives a good representation of the degree of nebulosity. Nebulosity is the fraction of the sky which is covered by cloud. This concept is not only descriptive but is related to the insolation fraction which is calculable insofar that one knows the hours of sunshine recorded on a heliograph and which is given by the expression

$$\sigma = \frac{SS}{SS_0} = \frac{\text{Sunshine hours}}{\text{Maximum possible sunshine hours}}$$

Statistical studies of the data show that the total radiation G on a surface can be related to diffuse radiation D and to the insolation fraction σ. As an example we give this relation

$$\frac{D}{G} = 0.9 - 0.8\sigma \text{(If } 0.15 < \sigma < 0.7) \tag{25}$$

or relation (10).

As we know G we can calculate the diffuse radiation D.

In practice, attention must be paid to average sunshine hours which are very important to any project.

The study of the insolation data at a weather station produces for each month the number of days on which insolation was above or below this average.

A consideration of these deviations allows an estimation of the storage which is required (see Section 4.6).

As we know σ we can correctly evaluate the energy received on a plane of any inclination. Generally we can determine from the data the energy received per day under clear sky conditions, G_0. If, for example, we choose a day in the middle of winter (15th February) then to find the energy received during the month of February we must

(a) calculate the total energy for the day

$$Q = \frac{2}{\pi} G \Delta H, \text{ where } G = \text{radiation at local noon and } \Delta H = \text{length of the day,}$$

(b) calculate the total energy for the month, assuming that the sky remains clear

$$Q_T = \sum_i^{28} Q_i$$

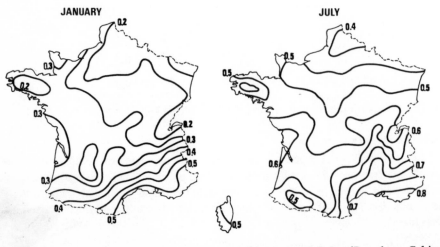

Figure 28. Average insolation fraction for January and July. (Based on Cahier de l'A.F.E.D.E.S., No. 1, adjusted)

JANUARY

Figure 29. Average sunshine hours for
January. (Based on Cahier de
l'A.F.E.D.E.S., No. 1, adjusted)

(c) take into account local meteorological data using σ to evaluate the energy
actually collected

$$Q_R = Q_T \times \sqrt{\sigma}$$

Another estimate can be made using the expressions in Paragraph 2.3.3
For example: at local noon the direct flux is 950 W m^{-2}:

$G_N = G \cos i$ at local noon $(h = \dfrac{\pi}{2} - \phi + \delta)$ and since $i = \dfrac{\pi}{2} - h + \alpha$ we have

$i = \phi - \delta - \alpha.$
Taking $\phi = 45°$ $\quad \delta = 15° \rightarrow h = 60°$ $\quad \Delta H = 11h$ $\quad \sigma = 0.4$
$G_N = G \cos(\phi - \delta - \alpha) = 950 \times \cos(-30) = 822$ W m^{-2}
$Q_{\text{for the day}} = \dfrac{2}{\pi} \times 11h \times 822 = 5{,}759$ Wh m^{-2}

$Q_{\text{for the day, actual}} = 5{,}759 \times \sqrt{\sigma} = 4{,}072$ Wh m^{-2}.

2.6.8 Air temperature

Air temperature variations are recorded on paper in a small covered shelter at a
standard 1.20 m. above the ground. The charts show the temperature every three
hours. It is therefore possible to calculate the degree-days necessary for estimating
heat losses. See later for a French map (Section 3.2.7, Figure 38).

Monthly charts are made (and thus an annual chart) reporting the average of the
minima (T_n), the average of the maxima (T_x) and the average of the two
$(T_n + T_x)/2$.

JANUARY, 1975

Station	Average of Min T_n	Average of Max T_x	$\frac{T_n+T_x}{2}$	Deviation from the norm.	ABSOLUTE MINIMUM Value	ABSOLUTE MINIMUM Date	ABSOLUTE MAXIMUM Value	ABSOLUTE MAXIMUM Date	Frost under cover	$T \le 0°0$	$0°0 < T \le 10°0$	$10°0 < T < 20°0$	$T \ge 20°0$	Average A 0h	3h	6h	9h	12h	15h	18h	21h
GROUP I																					
Dunkerque	5.5	9.8	7.7	3.8	0.8	28	14.0	15	0	0	29	2	0		7.5	7.5	7.6	8.5	8.6	8.3	7.9
Boulogne sur Mer	5.8	9.8	7.8	3.9	0.2	28	15.0	15	0	0	29	2	0		7.5	7.7	8.0	8.0	8.3	8.1	7.4
Le Touquet	5.7	10.1	7.9	4.1	1.7	5/	15.6	15	0	0	29	2	0	7.8	7.9	7.9	7.7	8.7	8.8	8.3	6.9
Abbeville	5.0	9.4	7.2	4.1	1.8	20	14.3	15	0	0	29	2	0	6.5	6.5	6.6	6.9	8.3	8.2	7.3	6.7
Lille	4.5	9.4	6.9	4.4	0.5	28	13.7	14	0	0	29	2	0	6.3	6.0	6.0	6.5	8.1	8.2	7.2	
Saint Quentin	3.9	8.7	6.3	4.3	0.1	26	13.6	15	0	0	30	1	0	5.5	5.3	5.3	5.7	7.5	7.5	6.5	5.9
GROUP II																					
Reims	3.8	8.9	6.4	4.4	-1.1	20	14.8	15	3	0	28	3	0	5.8	5.5	5.3	5.8	7.6	7.9	6.7	6.4
Romilly	3.1	9.1	6.1	3.8	-3.0	20	15.1	15	6	0	28	3	0	5.4	4.8	4.7	5.4	7.7	7.9	6.3	5.9
Auxerre	3.4	8.9	6.2	3.6	-2.4	2	15.6	15	7	1	27	3	0	5.3	4.9	4.6	5.0	7.1	7.8	6.3	5.9
Château Chinon	1.9	6.9	4.4	3.4	-3.7	4	12.6	13	10	4	26	1	0	3.6	3.3	3.3	4.1	5.2	5.1	4.3	3.9
Langres	1.8	6.1	4.0	3.6	-2.8	7	13.1	15	10	2	27	2	0	3.4	3.1	2.9	3.2	4.7	4.9	4.1	3.6
Saint Dizier	3.7	8.9	6.3	4.2	-3.0	20	16.2	15	3	0	27	4	0	5.6	5.3	5.0	5.6	7.8	7.7	6.5	6.0
GROUP III																					
Metz	2.9	8.6	5.8	4.6	-2.8	9	15.7	15	5	0	29	2	0	4.8	4.7	4.6	5.1	6.9	7.4	6.1	5.3
Nancy	3.1	8.1	5.6	4.7	-1.9	9/	14.6	15	4	0	30	1	0	4.8	4.6	4.7	5.1	6.8	7.4	6.1	5.3
Phalsbourg	2.3	7.3	4.8	4.9	-2.3	9	12.6	14	4	0	31	0	0	4.3	3.9	3.8	4.4	5.9	5.6	5.0	4.8
Strasbourg	2.0	8.2	5.1	4.6	-3.0	9	12.9	30	7	0	31	0	0	4.0	3.6	3.6	4.2	6.8	7.0	5.5	4.6
Mulhouse Bale	1.7	8.1	4.9	4.5	-2.2	5	13.8	17	10	1	30	0	0	3.4	3.3	3.1	4.0	6.7	6.8	4.8	4.1
GROUP IV																					
Belfort	1.8	6.9	4.4	4.3	-3.0	10	12.3	17	8	3	28	0	0	3.4	3.1	2.9	3.5	5.5	6.0	4.5	3.8
Luxeuil	1.2	8.2	4.7	4.3	-1.8	14	13.6	15	10	0	31	0	0	3.3	3.0	3.0	3.7	6.7	6.9	4.9	3.9
Besançon	2.3	8.1	5.2	3.9	-1.4	4	15.7	15	7	2	29	0	0	4.4	3.9	3.8	4.3	6.2	6.8	5.3	4.7
Dijon	2.9	7.8	5.4	4.0	-1.4	20	14.8	15	4	0	28	3	0	4.5	4.3	4.2	4.5	6.6	6.8	5.7	4.6
Mont St-Vincent	1.6	6.6	4.1	3.5	-4.6	3	13.1	15	12	4	25	2	0	3.2	3.1	3.3	3.7	5.2	4.9	4.1	3.6
Mâcon	3.0	8.8	5.9	4.1	-1.5	11	15.5	15	6	0	28	3	0	5.2	4.8	4.5	4.9	6.9	7.3	6.0	5.3
Les Sauvages	0.8	6.9	3.9	3.5	-7.0	2	12.1	31	12	6	25	0	0	2.9	2.8	2.5	3.4	5.5	5.1	3.8	3.5
Ambérieu	2.4	9.0	5.7	4.2	-4.1	10	16.8	15	9	3	25	3	0	5.1	4.6	4.9	5.3	7.7	7.9	6.5	6.0
Lyon	3.1	9.1	6.1	3.9	-1.8	3	15.8	15	3	1	27	3	0	5.1	5.0	4.7	5.9	7.8	8.1	6.7	6.1
Grenoble/Geoirs	1.0	8.4	4.7	3.5	-2.8	2	16.5	2	15	4	25	2	0	3.5	3.4	3.1	4.0	6.9	7.2	4.9	4.2
GROUP V																					

1975 (published by la Météorologie Nationale)

Figure 31. Temerature Chart (°C), averages of minima
T_n for January

2.6.9 Automatic processing of meteorological data

A major problem is to determine accurately the solar energy received on surfaces
with different inclinations.

We have seen the expressions involved (Section 2.5.5) and they show us that the
received energy depends upon:

the altitude of the sun h, the inclination of the surface i, and the orientation with
respect to the south and to the azimuth of the sun.

As h depends upon the latitude, the hour angle (HA) and the season (δ), in order to
know the energy received day by day, detailed calculations are necessary and it is
advantageous to analyse the meteorological data on a computer.

Method

(a) *Forming a data base.*   From the tables of total radiation, which are given for
each month, one can establish an average climatic year for the site by using all the
radiation measurements for a given station. Example: if one uses readings taken
over five years then one can calculate the hour by hour average for these five years
(using the figures at the bottom of the table). On a second pass one selects for each
month those data which best match the appropriate average. Thus the average year
is calculated.

Next one produces the data base by taking each month of the average year. The results are then transferred onto punched cards.

(b) *Simplified program flow chart.*

Meteorological data file

↓

Calculation of received energy
on a general inclined surface

↓

Loop on days of the month   Do J = 1,31

↓

Loop on hours   Do I = 1,14(5 to 19h)

↓

Print-out of daily results — total radiation G
diffuse radiation D
direct radiation I

↓

Selection of energy values
above a certain threshold

↓

Print-out of the results tables

(c) *Aims.* Practical applications, such as the heating of buildings or the production of domestic hot water using flat plate collectors, can only function when the collected energy exceeds a certain level. Therefore, it is very important to know for each month the number of hours during which the solar energy received by an inclined or vertical surface exceeds 1,000 W m^{-2}, 900 W m^{-2}, etc.

The Meteorological Office draws up statistics which are of great importance on this point; they furnish the following data for the stations recording total radiation G:

(1) The average frequencies (in $^o/_{oo}$) of daily sunshine hours at or above a given threshold (from 0 to 16 hours),

(2) for each month, the average daily frequencies (in $^o/_{oo}$) of the amounts of total radiation received on a horizontal surface, equal to or above a chosen threshold (from 0 to 8.3 kWh m^{-2}),

(3) the average frequencies, on an hourly basis, (from 5.00 h to 19.00h) for a given month in the year of reference (in $^o/_{oo}$) of the hourly amounts of total radiation received on a horizontal surface equal to or above a chosen threshold (from 0 to 8.3 kWh m^{-2}),

(4) the average frequencies (in $^o/_{oo}$) of the total radiation received on a horizontal surface over three consecutive days, equal to or greater than a chosen threshold (from 0 to 22 kWh m^{-2}).

NOTE: In practice, the collectors are always inclined at an angle i in order to be normal to the direct radiation during the chosen period and so the data must be subjected to an appropriate transformation by using the formulae given in paragraph 2.5.5. This can be done on a pocket calculator.

The same type of analysis can be done with temperatures if one wishes to compare solar gains on façades or collectors with losses which depend on the external temperature.

One can thus determine for the most unfavourable periods the magnitudes of the losses in relation to the inputs and can estimate the degree of supplementary heating and amount of storage required. We will return to the energy balance for a solar house in the next chapter.

2.7 LOCAL GEOGRAPHY

Planning in flat open country generally presents no difficulties other than those due to such obstacles as buildings, vegetation, etc.

Regions of pronounced elevations and depressions (mountains, certain coastal sites, undulating areas) can, however, produce losses in solar gains, as will be shown by an example.

The method consists in transferring to a graph the height and azimuth of the sun

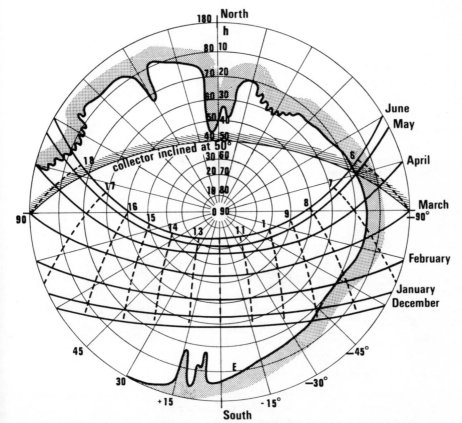

Figure 32. Example of a site survey made at Capri, Italy. Screening caused by relief features

and a relief map of the site under consideration. This can be done by using the same unit of measurement (angular value) for all of the elevations involved.

Thus, for unfavoured periods there will be radiation gaps caused by any screens.

The graph shown in Figure 32 shows that one loses about 1½ hours (or 18 to 23% of the radiation) which correspondingly delays the rise in temperature of a collector using water as the heat transfer medium.

2.7.1 Influence of the surroundings

The main factors here are buildings and vegetation, which can throw shadows onto the collector surfaces and thus impair efficient operation of the installation.

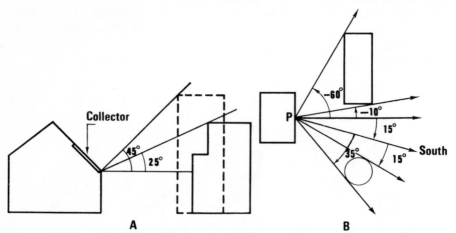

Figure 33. Plan and elevation showing different angles. An instance in which the building is rotated through 15° from the south

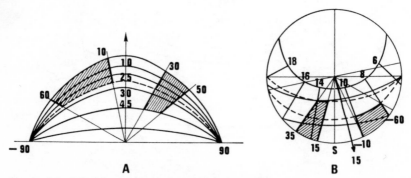

Figure 34. Sunshine curve and physical obstacles

A. Comparison of the shadow curves with the height of the buildings (25° and 45°) for the site shown in Figure 33.
B. Superposition of the sunshine graph for the latitude under consideration with the location of buildings on the ground (cf. plan-view Figure 33) and their respective heights (25° and 45°)

Two factors should be considered:

the height of an obstacle and the shadow it casts (the shadow cast varies with the seasons), and
the ground area affected by the cast shadow as it sweeps around during the day.

Two categories of angle must, therefore, be determined:

the height of each obstacle, and
the angle through which, at the collectors, the edges of the obstacles can be seen (the azimuth).

These observations are summarized in Figures 33 and 34, which show how the curves of insolation can be determined from the solar diagram.

3
Thermal Data

3.1 INTRODUCTION

3.1.1 Buildings and thermal exchanges

From a strictly climatic point of view, the walls of a dwelling separate two environments: the external environment, which is uncontrollable and subject to climatic factors, and the internal environment, which is theoretically controllable and has an artificial climate dependant upon local factors (external temperature, humidity, insolation, wind, rainfall), with a level of thermal comfort which provides the best possible conditions for the occupants.

In our temperate latitudes, the large variations in temperature necessitate a different approach for winter and summer.

A building contains within its shell (vertical, inclined, and horizontal surfaces) a certain volume of air which is to be maintained in thermal equilibrium at a given temperature.

The shell formed by the walls, roof, and floor can be made of traditional materials (stone, wood, clay) or prefabricated materials (glass, aluminium, concrete, iron, etc.). Each of these materials has a different internal composition (amorphous crystalline for glass, heterogeneous for concrete, etc.) which gives them a solid structure but different qualities.

3.1.2 The shape of the building shell

The shell limits the habitable volume and its optimal shape will depend upon the local external climatic conditions. For example, an igloo of hemispherical shape is

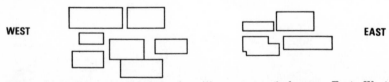

Figure 35. Schematic drawing of a village arranged along an East—West axis, permitting radiation on only one face. This grouping of houses gives effective protection against radiation

44

adapted to reduce the effects of wind and thermal losses. Arabic houses which are rectangular with few openings, and in close proximity to each other in order to avoid the maximum solar radiation, are usually oriented on an East—West axis and coloured white to promote reflection of radiation.

3.1.3 Materials

By their physical qualities different materials are capable of reflecting or collecting solar radiation and of transmitting or attenuating thermal variation to a greater or lesser extent.

These physical qualities can be measured and in consequence the materials can be classified. In particular, the thermal inertia of a building is a deciding factor in a warm dry climate. There are other problems, such as ventilation, condensation, etc.

In modern wall construction, one customarily distributes each of these functions (load-bearing, insulation, condensation, water-tightness) among the different constituent components.

In the past, wall materials fulfilled these multiple functions with varying efficiency because a building was dependant upon local materials. For example, the basalt stone of the Auvergne provides very little insulation at low winter temperatures.

Finally, since 1974 the energy problem has assumed a considerable economic importance and this has led governments to impose exact norms as to the degree of thermal insulation of dwellings.

3.2 INSULATION – HEAT LOSSES FROM BUILDINGS

Heat transfer occurs through conduction, convection, and radiation. Let us consider a wall made of three homogeneous materials with thermal conductivities of $\lambda_1, \lambda_2, \lambda_3$ and with thicknesses of e_1, e_2, e_3.

3.2.1. Convection and radiation from an internal surface

In still air the convection coefficient is given by

$\alpha_c = 4.6$ W m^{-2} °C^{-1} for a vertical surface,
$\alpha_c = 11.6$ W m^{-2} °C^{-1} for a horizontal surface, with rising air flow.

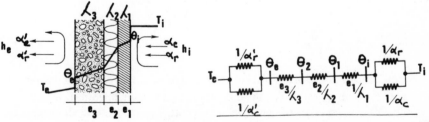

Figure 36. Diagram of temperatures in a composite wall and its electrical analogue

The radiation coefficient α_r has the value 4.6 W m^{-2} °C^{-1} for the majority of materials used in building (except for aluminium).

These two coefficients can be combined in the single term h_i, called the surface heat exchange coefficient for the internal wall.

$$h_i = \alpha_r + \alpha_c = 9.3 \text{ W m}^{-2} \text{ °C}^{-1} \text{ (8 kcal m}^{-2} \text{ h}^{-1} \text{ °C}^{-1}\text{)}.$$

For a vertical surface the heat flux can be written:

$$q = h_i(T_i - \theta_i).$$

3.2.2 Conduction through the three layers

Layer 1: $q_1 = \dfrac{\lambda_1}{e_1}(\theta_i - \theta_1),$

Layer 2: $q_2 = \dfrac{\lambda_2}{e_2}(\theta_1 - \theta_2),$ \hfill (26)

Layer 3: $q_3 = \dfrac{\lambda_3}{e_3}(\theta_2 - \theta_e).$

3.2.3 Convection and radiation on the external surface

One can define a surface heat exchange coefficient for the external wall by h_e,

$$h_e = \alpha'_c + \alpha'_r \tag{27}$$

with $\alpha'_c = 9.9$ W m^{-2} °C^{-1} for a vertical surface

$\alpha'_c = 11.9$ W m^{-2} °C^{-1} for a horizontal surface

$\alpha'_r = 4.6$ W m^{-2} °C^{-1} for a vertical surface

The heat flux can therefore be written:

$$q = \frac{T_i - T_e}{\dfrac{1}{h_i} + \sum\limits_{i=1}^{3} \dfrac{e_i}{\lambda_i} + \dfrac{1}{h_e}} \tag{28}$$

and one can write

$$\frac{1}{K} = \frac{1}{h_i} + \sum\limits_{i=1}^{3} \frac{e_i}{\lambda_i} + \frac{1}{h_e}. \tag{29}$$

The conductivities λ_i are given in the standard tables or by the manufacturers. K is the total coefficient of thermal transmission for a given wall, in units of W m^{-2} °C^{-1}.

In a building, two types of coefficient K occur: K_{op} for opaque elements (walls, floors, roofs), and K_{glass} for the windows.

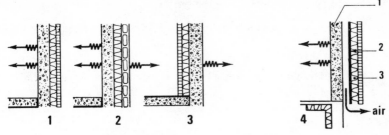

Figure 37. Different types of insulation.

1. Exterior insulation, high inertia.
2. Insulation within the wall.
3. Interior insulation, low inertia.

4. Exterior insulation with a cavity.
   Small 1. inner wall,
   Small 2. aluminium,
   Small 3. insulation

3.2.4 Ventilation

For hygiene reasons and because of water vapour it is necessary to change the air at a rate not less than one air change per room per hour. This incurs a heat loss: q.

$$q = 0.34\, NV(T_i - T_e) \tag{30}$$

where 0.34 represents the heat capacity of 1 m^3 of air at 20°C, V is the volume of the building and N is the hourly air change rate ($N = 1$ for living rooms).

3.2.5 Total heat losses from a building: the coefficient G.

The walls form a shell which contains a volume of air V at an ambient temperature T_i. The external environment contains masses of air at a temperature T_e. It is, therefore, possible to write a thermal balance sheet which allows for all of the above factors (opaque walls of area S_{op}, glazing of area S_g, ventilation).

$$q = \Sigma S_{op} K_{op}(T_i - T_e) + \Sigma S_{glass} K_{glass}(T_i - T_e) + 0.34\, NV(T_i - T_e) \tag{31}$$

On dividing each of these terms by V (the internal volume) we get

$$G = \frac{\Sigma S_{op} K_{op}}{V} + \frac{\Sigma S_{glass} K_{glass}}{V} + 0.34\, N \text{ in W m}^{-3}\,°C^{-1} \tag{32}$$

In most countries some recommendations or regulations have been instituted on the maximum value of G considered acceptable. This varies according to the building type, and in the French regulations of 18th April, 1974, we find

Values of G (W m^{-3} °C^{-1})			
	Winter Climatic Zone		
House Type	A	B	C
Volume < 150 m^3	1.60	1.75	2.80
150 m$^3 \leqslant V < 300$ m^3	1.45	1.60	1.90
$V \geqslant 300$ m^3	1.30	1.45	1.75

3.2.6 Losses at a given site over the winter period

Equation 31 can be expressed in the following form to produce a daily figure,

$$q = GV(T_i - T_e) \times 24 \text{ h}$$

In order to determine the losses during the entire winter period, we must introduce degree-days, that is, to measure, day by day, the variation between the internal ambient temperature taken at $18°$ and the external temperature which is taken at hourly intervals.

On dividing by 24 this then produces the result for the day and one obtains

$$D = \sum_{\substack{i = 1st \\ \text{October}}}^{31 \text{ March}} (T_i - T_e) \text{ degree-days} \tag{33}$$

NOTE: Degree-days D can be calculated for different ambient temperatures T_i and for different periods.

3.2.7 Losses during the winter period

$$Q = GV\, 24 \text{ h } D \tag{34}$$

Values for D can be calculated for the 100 meteorological stations which record temperature. A map for France can thus be drawn up with degree-day values (D).

NOTE: the calculations giving G assume an equilibrium situation. In actual fact, the external temperature varies throughout the day; there are energy inputs to the east, south, and west walls and diffuse radiation on the north wall which are not accounted for in the calculations.

A correct approach to the problem assumes the treatment of a non-equilibrium system which takes account of the temperature variations and the energy inputs as a function of time.

The expression can be written as

$$Q(t) = GV(T_i - T_e(t)) - \Sigma A(t) \tag{35}$$

where $A(t)$ represents the solar gains. Taking them into acccount reduces the calculated heat loss by approximately 12% in winter.

These solar gains occur primarily because of the windows. For a realistic installation which is economic when compared to conventional solutions it is necessary to follow the daily gains and losses in order to predict the supplementary heating and the storage required for the least favourable conditions.

Supplementary heating

Only in exceptional cases can heating be ensured solely by solar energy due to high costs (the cost of installation, excessive surface area of collector, architectural constraints, etc.). It is, therefore, necessary to instal a conventional heating system (wood, oil, gas, or electricity).

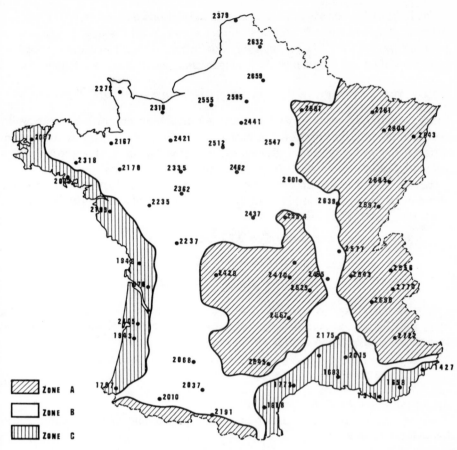

Figure 38. Map of degree-days for the period 1st October to 31st March. Internal temperature of 18 °C. A, B, and C are the three climatic zones.

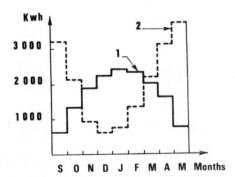

Figure 39. Diagram showing monthly heat gains and heat demand.

1. Demand.
2. Collected energy

With Equation 35 we can improve the calculations derived from the meteorological data and prepare monthly statistics. For a daily study a computer is essential, but for the ten-day or monthly calculations manual calculation is possible.

A monthly graph thus shows the heat gains and heat demand (see Figure 39).

3.3 HEAT STORED IN BUILDING STRUCTURE – THERMAL INERTIA

Solar gains through glazed openings cause a temperature rise in the interior walls, the flooring and the furniture. In effect, 85% of incident energy is transmitted through the ordinary 3 mm glazing which forms the shell of a room, with a reduction for reflection which depends on the angle of incidence.

Energy is exchanged in the mass of materials by conduction and with the ambient air by convection. The storage of a quantity of heat depends upon the thermal capacity of the materials ρCV.

For concrete, brick, stone $C \simeq 0.22$ kcal kg^{-1} $^{\circ}$C^{-1}

water $\qquad\qquad\qquad C \simeq 1$ kcal kg^{-1} $^{\circ}$C^{-1}

air $\qquad\qquad\qquad\quad C \simeq 0.24$ kcal kg^{-1} $^{\circ}$C^{-1}

The stored energy is transferred to the surroundings with a delay depending upon the thermal properties of the construction materials. From heat theory we have the general equation of heat propagation:

$$\frac{\partial^2 \theta}{\partial t^2} = \frac{\rho c}{\lambda} \frac{\partial \theta}{\partial t} \tag{36}$$

and we can let $a = \lambda/\rho C$, the thermal diffusivity.

In the case of a periodic variation of period T, the temperature at distance x into the wall can be written

$$\theta = \mu \theta_0 \frac{\sin 2\pi(t - \phi)}{T} \tag{37}$$

where $\mu(x)$ is the attenuation, and $\phi(x)$ is the delay.

If one substitutes this expression in Equation (36) we get

$$\mu = \exp\left(-\sqrt{\frac{\pi}{aT}} x\right) = \exp - \left(\sqrt{\frac{\pi\lambda\rho C}{T}} \frac{x}{\lambda}\right) \tag{38}$$

$$\phi = \frac{T}{2} \sqrt{\frac{1}{\pi aT}} x = \frac{T}{2} \sqrt{\frac{\lambda\rho C}{\pi T}} \frac{x}{\lambda}$$

The two coefficients depend upon $\sqrt{\lambda\rho C}$ (where $\lambda\rho C$ is the admissivity). Let us look for example at its value for various materials.

	λ kcal m^{-2} $^{\circ}$C^{-1}	C	ρ kg m^{-3}	$\lambda\rho C$	$\sqrt{\lambda\rho C}$
light insulating materials	0.04	0.22	200	18	1.35
light concrete	0.3	0.22	800	53	7.3
limestone	0.9	0.22	1,700	340	18.5

When comparing three walls which have the same thermal resistance, e.g. $K = 1.4 \text{ kcal m}^{-2} \text{ h}^{-1} \text{ }^\circ\text{C}^{-1}$ one can easily show the thicknesses of wall needed for different materials, and using Equation 38 show the different thermal behaviour (taking $T = 24$ h).

Table giving the attenuation and phase shift of a thermal wave passing through a wall

	Thickness	Attenuation	Phase delay
Insulation	2 cm	0.67	1 h
Light concrete	15 cm	0.12	5 h
Limestone	45 cm	0.005	13 h

This explains why traditional buildings ensure thermal comfort in summer much more efficiently than modern constructions.

If we examine Figure 40, the difference between Q_1 and Q_2 shows the heat losses by conduction to the outside, since in winter: $t_{ext} < t_{int}$.

The manner in which the heat is released depends upon the wall and the position of the insulation (internal, inside the wall, or external). Positioning the insulation on the outside is best for storing the heat arising from solar gains through the glazed areas and, further, limits losses to the outside. The problems posed by this technique are not yet solved although the methods exist. Internal insulation limits the absorption of the radiation.

Wall temperature and comfort

The external gains collected raise the temperature of the different receiving walls and influence the ambient temperature.

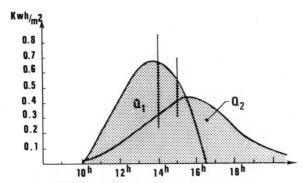

Figure 40. Instantaneous and real gains in a traditional building.

Q_1 heat gained from solar radiation through glazed openings.
Q_2 heat lost to the outside

It can be shown that the ambient temperature of a building is $t_{\text{ambient}} = (t_{\text{air}} + \theta_r)/2$, for still internal air, where θ_r is the radiant temperature:

$$\theta_r = f_1 t_1 + f_2 t_2 + f_3 t_3 + f_4 t_4 \text{ etc.,}$$

where t_1, t_2, t_3, t_4 are the temperatures of the different walls and f_1, f_2, f_3, f_4 are the factors which express the solid angle through which, for a given point on the building, one sees the wall under consideration.

Conclusion   Any excessive elevation of wall temperature causes discomfort by its effect on the internal temperature — a limiting factor in summer for light structures — but it is very important in winter and should be considered in a solar dwelling with as much care as the heat exchange surfaces which influence air temperature.

A solar house

The approach to the thermal optimization of a solar house is different from that for a dwelling with all-electric heating. In the latter case, because of the high price per kilowatt hour, it is important to avoid heat losses by providing good thermal insulation. Current experiments on solar housing show that thermal insulation is equally important here so as to minimize collector areas and keep costs within tolerable limits.

Clearly it would appear that the solar house must be considered as a solar collector in its entirety. The lay-out and uses of the rooms must be designed as a function of orientation and sunshine.

A study of the functions of occupation at different times (for example, a bedroom at the east would greatly benefit from morning radiation, etc.) illustrates this.

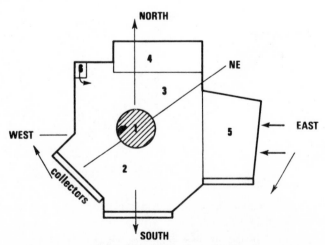

Figure 41. Plan-view of a solar house

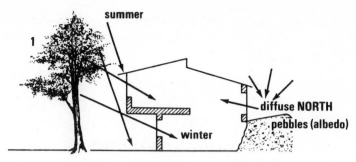

Figure 42. Elevation of a solar house

The creation of buffer zones (for example, garage and storerooms along the north wall, and greenhouses and conservatories on other faces) helps to keep the walls relatively warm.

The study of losses under variable weather conditions shows the approximate nature of real control systems which assume near-stable conditions.

Sunshine is a non-predictable variable; stable conditions cannot be guaranteed to any building wholly dependant upon it.

4

Design Concepts

Every system for the thermal conversion of solar energy includes the following elements:

a solar collector surface,
a heat transfer system which transfers the energy extracted by the collector to the thermal store,
a thermal store,
a distribution system.

There are two types of collector: non-concentrating and concentrating.

4.1 THE FLAT PLATE COLLECTOR AND ITS OPERATION

All bodies exposed to the sun heat up by the transformation of the received radiant energy into heat energy. The heat energy thus created is transferred to a heat transfer fluid. In these devices the temperature rarely exceeds 100 °C (usually it is 60 °C or less).

Suitable heat transfer fluids are: water, air, and refrigerant fluids in either the liquid or vapour phase; oil or antifreeze. Flat plate collectors generally use water or air.

4.1.1 Construction of a collector plate

A flat plate collector is a box consisting of three components:
a front part, the collector itself, which is exposed to radiation,

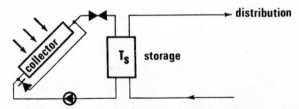

Figure 43. Schematic diagram of an installation

54

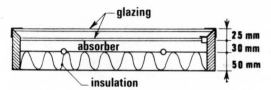

Figure 44. Typical cross-section of a collector plate

the absorber where thermal conversion takes place and the backing.

The front section should:

(1) receive radiation without reflection, diffusion, or unnecessary absorption,
(2) oppose thermal losses through convection and radiation since the absorber is raised to a temperature higher than that of the surroundings.

This dual function is realized by placing a transparent cover (glass or plastic) at a small distance above the absorber.

4.1.2 Radiation losses

The glass or plastic coverplates let through the incident solar radiation (function 1) with a high transmission factor τ (about 0.85), but they are opaque to the infrared radiation emitted by the absorber which has been raised to a temperature between $35°$ and $100\,°C$ (wavelength $>20\,\mu$).

This is called the greenhouse effect. The internal face of the covering absorbs infrared radiation, undergoes a rise in temperature and in its turn radiates half towards the outside and half towards the absorber.

In the thermal balance sheet radiation losses are reduced by half. It would be possible to further reduce these losses by using several layers of glazing but in practice double glazing is not exceeded for two reasons:

(a) reduction of the transmission factor ($\tau = 0.65$ for double glazing)
(b) increased cost.

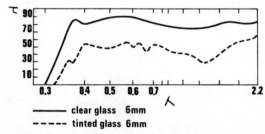

Figure 45. Transmission characteristics through glazing

Convection losses

The transparent cover also limits losses by convection. Air is one of the better insulators and a thin layer will insulate well provided it is not thicker than 3 cm. Above this convection currents appear. (Polystyrene owes much of its high insulation properties to the bubbles of air trapped within it which are unable to move.) In practice, 3 cm of air is generally used between the absorber and the cover so as to limit heat exchange by conduction through the stationary air. Pyrex glass and glass with regular surface corrugations can offer a benefit under low sun conditions (near-grazing incidence) by encouraging transmission.

Suitable plastic materials are: Therfane, Mylar, Altuglass.

4.1.3 Properties of glazing and of double glazing

The spectral distribution of natural daylight is:
1% ultraviolet, 53% visible, and 46% infrared.

		Solar energy factors			Overall factor
	Thickness	reflection	absorption	transmission	
Drawn glass	3 mm	0.07	0.06	0.87	0.88
Polyglass	6 mm	0.13	0.20	0.67	0.72
Terphane		0.12	0.03	0.85	
Mylar		0.12	0.03	0.85	

In order to limit losses through the cover it can be treated with a selective coating on its internal face which preferentially reflects the infrared radiation coming from the absorber (high cost).

4.1.4 The absorber

This is very often a metal plate of good thermal conductivity on which has been deposited a surface coating which is absorbant to solar radiation.

The simplest consists of painting the plate matt black (absorption coefficient = 0.85 to 0.9). In water collectors the absorber also acts as the heat exchange surface. For example, panel radiators, painted black, make inexpensive absorbers while showing good efficiency.

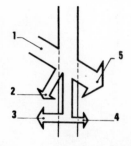

Figure 45a. Components of flux.

1. Incident solar flux (100%).
2. Reflected component flux ($r\%$).
3. Flux absorbed and rejected towards the outside.
4. Flux absorbed and rejected towards the interior of the building.
5. Flux transmitted (energy transmission factor τ).
4 + 5 = Solar energy gain

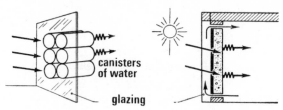

Figure 46. Examples of absorbing wall and thermal store

In a similar manner, any surface coated in matt black paint can serve as a collector and black painted walls will provide two additional functions: structural, and thermal storage in the mass.

In particular, canisters filled with water, or a concrete or stone wall of a certain thickness can play the part of an absorbant wall and thermal store (cf. paragraph 5.2.1).

Improving the efficiency of the absorber

In order to reduce losses from the collector and increase its efficiency, the absorber surface can be coated with a selective coating which has the highest possible absorptivity α for radiation of solar wavelengths (about $2\,\mu$) and the smallest possible emissivity ϵ for the infrared radiation emitted by the absorber (wavelengths greater than $20\,\mu$).

In practice a chemical is deposited on the absorber which allows a value $\alpha/\epsilon = 10$.

Examples of selective coatings

There are four methods:

(a) a multilayer coating: obtained by depositing several layers of material in different thicknesses,

(b) treatment of the collector material: usually treatment of a copper plate to produce copper oxides with a ratio $\alpha/\epsilon = 6$,

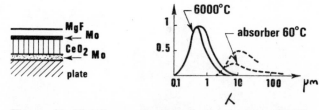

Figure 47. Multilayer selective coating and comparison between the curve of solar radiation and that of the absorber

58

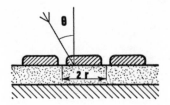

Figure 48. Globular deposits

(c) mesh: a thin strip of metal placed on a supporting grid basically transparent to solar radiation,
(d) globular deposits.

Globular films are formed of electrically insulated metallic particles which are very close to each other and are transparent to infrared radiation.

Methods (c) and (d) are still at the experimental stage. Their benefit lies in a ratio $\alpha/\epsilon > 10$.

In practice, the use of selective surfaces is justified for temperatures over 60 °C.

Another means of reducing losses by convection and radiation from the absorber surface.

Honeycomb-like cellular structures are built up on the absorber surface.

These structures immobilize the air inside the cells and limit radiation. Their main difficulty lies in the development of an economical material which is resistant to high temperatures and ultraviolet radiation. In conventional flat plate collectors this technique is not used

4.1.5 The backing

This is generally made up of an insulating layer of 5 to 10 cm thick (glass fibre or expanded plastic foam) which reduces losses by convection and thus prevents the collector surface away from the sun from cooling down.

There are cases where the backing can serve another function besides thermal (as an element of roofing, loadbearing wall, or infill).

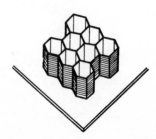

Figure 49. Cellular structure

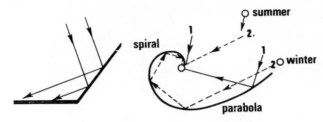

Figure 50. Two intensification devices.

1 Intensification;   2. concentration

4.1.6 Intensification techniques

In order to reduce thermal losses, one can also equip the collector with suitably oriented reflective surfaces which direct the incident radiation towards the absorber with a correspondingly smaller surface area.

4.1.7 Orientation and inclination of collectors

At our latitudes of the northern hemisphere the curves of received energy on a horizontal east, west or south face show that an orientation of due south, south-east, or south-west is favourable. When we remember that the solar declination δ varies from $-23°27'$ to $+23°27'$ and that $h_{max} = \pi/2 - \phi + \delta$ at local noon, various options are open to us for the inclination of the collector, depending on the optimization required. This can be

annual (production of hot water): $\alpha = \phi = $ latitude of the site ($\delta_{aver} = 0$ over the year);
winter (heating): $\alpha = \phi + 11°75$ ($\delta_{aver \, for \, winter} = -11°75$);
summer: $\alpha = \phi - 11°75$;

the two other main possibilities are

vertical (heating): $\alpha = 90°$,
horizontal: $\alpha = 0°$.

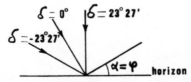

Figure 51. Inclination of collectors.
$$h_{max} + \alpha + \pi/2 = \pi$$
$$\alpha = \pi/2 - h_{max} = \phi - \delta$$

60

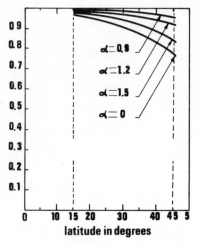

Figure 52. South facing surface inclined at an angle α to the horizontal. Variation of annual insolation fraction τ with latitude for different plate inclinations

4.1.8 The different functions of a solar collector

These can be separated into three classes: energy functions, architectural functions and aesthetic functions.

The first class contains two groupings:
(a) thermal functions: heating applications (dwellings, glasshouses), hot water production, distillation (fresh water production, salt-extraction), refrigeration by absorption processes,
(b) functions caused by photons of light: photovoltaic electricity production, photosynthesis, etc.

The second class can be divided into three groups:
(a) daylight,
(b) architectural integration: this arises through the addition of separate solar collectors to a group of buildings. It is important that these adaptations have minimal effect upon the skyline and to the style of the building. It is especially important not to break the harmony of a traditional dwelling, and the collectors in this case should be sited at the side, behind a mound or shrubs.
(c) architectural incorporation: the use of collectors as a functional wall.

Aesthetic functions (third category)

This is an important architectural problem and should not be underestimated: the dark surfaces of collectors can be intrusive if their siting has not been properly investigated. An example of architectural incorporation is a collector which produces heating and electricity and is also used as a roofing element.

4.1.9 Estimation of collector temperatures

Surface temperature of the absorber (black surface), unglazed

We will assume an absorptivity $\alpha = 0.95$. A surface at a temperature above that of the air will lose heat by convection and radiation. There is, therefore, an overall heat transfer h_e.

$$h_e = \alpha_c + \alpha_r,$$

where α_r is the radiation heat transfer coefficient and α_c is the convection heat transfer coefficient which depends upon airspeed.

$h_e = 9 + 3.8\,V$ (9 in W m^{-2} $^{\circ}$C^{-1}; V, airspeed in m s^{-1}).

At thermal equilibrium the heat collected = the heat lost.

$Q_r = Q_p$, that is $I\alpha = h_e\Delta t$

where Q_r = the absorbed power (W m^{-2})

Q_p = the dissipated power (W m^{-2})

I = radiation intensity (W m^{-2})

α = absorptivity

$\Delta t = t - t_{ambient}.$

If $V = 1$ m s^{-1}, $h_e = 12.8$ W m^{-2} $^{\circ}$C^{-1}, $I = 500$ W m^{-2}, $\alpha = 0.95$

$$\Delta t = \frac{500 \times 0.95}{12.8} = 37\,^{\circ}\text{C}$$

above air temperature (assuming that there is no heat loss through the back face of the absorber.

Temperature reached when including the thermal properties of the absorber

Incident heat collected = heat gain + heat loss;

therefore, $I = I\alpha h S = C\Delta t + h_e h S \Delta t$

where C = the thermal capacity of the absorber Wh $^{\circ}$C^{-1} (mc)

where m is the mass in kg and c = specific heat,

for $h = 1$ hour and $S = 2$ m^2

$$\Delta t = \frac{I \times \alpha \times 1 \times 2}{C + h_e \times 2}.$$

Let us consider a steel radiator of area 2 m^2 and weight 10 kg. ($c = 0.13$ Wh kg^{-1} $^{\circ}$C^{-1}), containing 4 litres of water ($c = 1.16$ Wh l^{-1} $^{\circ}$C^{-1}); and an incident intensity $I = 450$ W m^{-2}

$$C = 10 \times 0.13 + 4 \times 1.16 = 1.3 + 4.64 = 5.94 \text{ Wh } ^{\circ}\text{C}^{-1}$$

$$\Delta t = \frac{450 \times 0.95 \times 2}{5.94 + 12.8 \times 2} = \frac{855}{31.5} = 27\,^{\circ}\text{C}.$$

The calculation is made assuming a stationary fluid. C includes the effect of the component due to the fluid and shows that only small quantities of fluid per square metre are required if substantial result is to be obtained.

Temperature reached by the absorber with single or double glazing

Thermal losses are mainly those due to convection and conduction, and one can use an overall coefficient K for glass.

$K = 5.5 \text{ W m}^{-2}\,{}^{\circ}\text{C}$ for single glazing

$K = 3.3 \text{ W m}^{-2}\,{}^{\circ}\text{C}$ for double glazing.

For low temperatures the heat losses can be written

$$Q_p = K \Delta t h S.$$

On the other hand, the absorption by the glass has to be accounted for and $\tau = 0.88$ for single glazing; $\tau = 0.72$ for double glazing, and the received heat is

$$Q_r = I\alpha\tau S$$

$$Q_{rs} = 450 \times 0.95 \times 0.88 \times 2 = 376.20 \times 2 = 752.40 \text{ W (single glazing)}$$

$$Q_{rd} = 450 \times 0.95 \times 0.72 \times 2 = 307.80 \times 2 = 615.60 \text{ W (double glazing)}$$

and $\quad \Delta t = \dfrac{376.20 \times 2}{5.94 + 5.5 \times 2} = \dfrac{752.40}{16.9} = 44\,{}^{\circ}\text{C}$ for single glazing

$\quad \Delta t = \dfrac{307.80 \times 2}{5.94 + 3.3 \times 2} = \dfrac{615.60}{12.54} = 49\,{}^{\circ}\text{C}$ for double glazing.

Absorber temperature when the thermal fluid is moving (thermal-siphon or pumped circulation)

The fluid will be raised to a higher temperature than when it entered by means of a portion of the heat extracted from the absorber. The temperature of the absorber will be consequently at a lower level than the equilibrium temperature calculated above. This reduces the heat losses (since the temperature is lower).

Let us include in the thermal capacity C the quantity of water flowing during the time period considered.

Taking, for example, a flow rate $\dot{m}$ of 10 litres $\text{h}^{-1}\,\text{m}^{-2}$, $\dot{m}C_p$ is the thermal mass flow of the fluid and we have, taking a collector area of $S = 2 \text{ m}^2$

$$C = 10 \times 0.13 + 20 \times 1.16$$

whence $\Delta t = \dfrac{450 \times 0.95 \times 0.88 \times 2}{5.5 \times 2 + 23.2 + 1.3} = \dfrac{752.4}{35.5} = 21\,{}^{\circ}\text{C}$ for single glazing

$\quad \Delta t = \dfrac{615.6}{31.1} = 20\,{}^{\circ}\text{C}$ for double glazing.

For example, with an air temperature $t_a = 12\,{}^{\circ}\text{C}$ and with the fluid temperature starting at the same temperature, then after 1 hour 20 litres of fluid are produced at

$12\,{}^{\circ}\text{C} + 21\,{}^{\circ}\text{C} = 33\,{}^{\circ}\text{C}$ for single glazing

$12\,{}^{\circ}\text{C} + 20\,{}^{\circ}\text{C} = 32\,{}^{\circ}\text{C}$ for double glazing.

This represents the start of the cycle. Under normal operation distinction must be made between two different Δt.

(1) The heat extracted and transferred to the fluid:

$$\Delta t = t_s - t_e$$

where t_s is the fluid temperature on leaving the collector and t_e the temperature on entry.

This measures the increase in temperature of the fluid through the absorber.

$Q_u = (t_s - t_e)\dot{m}C_p$ where $\dot{m}C_p$ is the thermal mass flow.

(2) Heat lost:

$$\Delta t = \frac{t_s + t_e}{2} - t_a$$

$$Q_p = KS\left(\frac{t_s + t_e}{2} - t_a\right).$$

(3) Heat collected:

$$Q_r = IS\tau\alpha$$

At thermal equilibrium $Q_u = Q_r - Q_p$

$$\dot{m}C_p(t_s - t_e) = IS\tau\alpha - KS\left(\frac{t_s + t_e}{2} - t_a\right)$$

whence

$$t_s = \frac{IS\tau\alpha + t_e\left(\dot{m}C_p - \dfrac{SK}{2}\right) + KSt_a}{\dot{m}C_p + \dfrac{KS}{2}}$$

permitting us to determine the exit temperature from the collector if we know the other data.

Numerical Example: By taking the values given below one finds: $t_s = 54\,^\circ\text{C}$ ($I = 500$ W m^{-2}, $\tau = 0.88$, $\alpha = 0.95$, $K = 5.5$, $t_a = 0\,^\circ\text{C}$, $\dot{m}C_p = 10\,\text{l h}^{-1}$, $S = 2$ m^2, $t_e = 10\,^\circ\text{C}$).

4.1.10 Evaluation of collector losses

Coverplate

(a) Reflection of radiation: The reflection coefficient R reduces the incident energy by a factor $R = (n - 1)^2/(n + 1)^2$ under normal incidence (Fresnel's formula).

For ordinary glass $n = 1.5$, therefore

$$R = \left(\frac{1.5 - 1}{1.5 + 1}\right)^2 = \frac{1}{25} = 4\%.$$

There is a reflection from each face, leading to $R = 8\%$, but generally the practical value is a little less, 7%, for ordinary window panes.

For a plastic (e.g., Mylar), $n = 1.65$,

$$R = \left(\frac{1.65 - 1}{1.65 + 1}\right)^2 \frac{1}{1.65} = 6.1\%$$

which gives us 12% when both faces are taken into account.

(b) Radiation absorption: Its value is furnished by Lambert–Bouguer's law $I = I^* \exp(-Kx)$ where K is the absorption factor for glass and x is the thickness of the glass.

If K is taken as 0.04 cm^{-1} and $x = 3$ mm then this equation gives an absorption of about 7%, which is typical for glazings currently on the market.

Convection losses in the air layer between the absorber and the glazing:

$$q = \alpha_c S(t_{abs} - t_{glazing}).$$

In order to find α_c one uses the Nusselt and Grashof numbers from fluid mechanics.

Nusselt number:

$$N_u = \alpha_c/(\lambda/e)$$

Grashof number:

$$G_r = \frac{\rho^2 g \beta \Delta T}{\mu^2}$$

λ thermal conductivity of air, e thickness of air layer, g acceleration of gravity and μ viscosity.

$$\beta = \frac{1}{T(\text{degrees absolute})}$$

$\rho = $ kg m^{-3} density.

The relation between N_u and G_r is

$$N_u = 0.093(G_r)^{0.31} \qquad 10^4 < G_r^2 < 10^7$$

(for a collector inclined at 45° to the horizontal).

This leads to a value of $\alpha_c = 3.5$ W m^{-2} °C^{-1} and for $t_{abs} - t_{glazing} = 50$ °C

$$Q_c = 175 \, \text{W m}^{-2}.$$

Losses by radiation (absorber to glazing): These occur between two parallel surfaces and the quantity of heat exchanged can be written:

$$Q_r = \sigma S_{abs}(t_{abs}^4 - t_{glazing}^4) \frac{1}{(1/\epsilon_{abs} - 1/\epsilon_{glazing} - 1)} f(S_{abs}, S_{glazing})$$

$\epsilon_{abs} = 0.95$ and $\epsilon_{glazing} = 0.93$ — the emissivity of the two surfaces, $\sigma = $ Boltzmann constant (4×10^{-8} kcal h^{-1}K^{4}m^{-2})

$f(S_{abs}, S_{glazing})$: geometrical factor obtained from tables which give $f = 0.8$ for

$$t_{abs} = 70\,°C, t_{glazing} = 50\,°C, S_{abs} = 1\,m^2, Q = 123\,W\,m^{-2}.$$

Losses by conduction through the collector plate: The layer of glass fibre has a thermal conductivity $\lambda = 0.041\,W\,m^{-2}\,°C^{-1}$, for a thickness of 0.075 m:

$$Q_\lambda = \frac{0.041 \times 1}{0.075}\,(t_{abs} - t_i)$$

$t_{abs} = 70\,°C$ and $t_i = 20\,°C$, therefore

$$Q_\lambda = 27.33\,W\,m^{-2}.$$

It is necessary to include the losses through the lateral walls. Finally part of the radiation emitted by the absorber is absorbed by the glazing which heats up and then loses its heat to the outside air. Assessment of Losses: Let us take an incident flux of 1,000 W m^{-2}:

reflection: $1,000 \times 8\% =$ 80 W
absorption: $1,000 \times 7\% =$ 70 W
_____
150 W

transmitted flux (85%) = incident flux − (reflected flux + absorbed flux), i.e.,

i.e.,   850 W   =   1,000 W   −   150 W

absorbed flux (81%) when one includes an absorption coefficient of less than 1.

$$850\,W \times 0.95 = 807.5\,W,$$

the remaining 4% is reflected: part passes through the glazing, the rest is absorbed by the walls.

Convection flux and radiation between the absorber and glazing:

radiation: 12%
convection: 17%
conduction: 3 to 5%.

This leads to total losses of 475 to 500 W m^{-2} for an incident flux of 1,000 W m^{-2}.

The usable energy therefore represents

$$\frac{1,000 - 475}{1,000} = \frac{525}{1,000} = 52\%.$$

This, therefore, is the maximum theoretical efficiency.

4.1.11 Mathematical models expressing collector operation

The losses estimated above can be written in more general forms.

Absorbed power:

$$Q_a(\text{watts}) = G\, \overline{\tau\alpha} \tag{39}$$

where G is the total incident illumination (direct, diffuse, earth, atmosphere) and $\overline{\tau\alpha}$ the mean optical factor.

A mean value is used as a simplification and a calculation is not made at each instant for the direct and diffuse radiations which would require separate absorption and transmission coefficients. We also consider total illumination, that is, the normal incident radiation multiplied by the average optical factor $\overline{\tau\alpha}$.

Dissipated power:

$$Q_p = Q_{\text{front}} + Q_{\text{rear}}$$
$$Q_{\text{front}} = K_{\text{front}}(t_m - t_a)$$
$$Q_{\text{rear}} = K_{\text{rear}}(t_m - t_a)$$
$$Q_p = (K_{\text{front}} + K_{\text{rear}})(t_m - t_a) = K(t_m - t_a) \tag{40}$$

where $K, K_{\text{front}}, K_{\text{rear}}$ are the heat transfer coefficients (total, front and rear) for the absorber.

Rear Face:

$$K_{\text{rear}} = \frac{1}{e_i/\lambda_i + 1/h_{\text{rear}}} \tag{41}$$

where h_{rear} is the convection coefficient and e_i/λ_i the thermal resistance of the insulation.

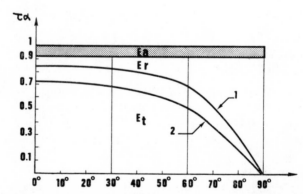

Figure 53. Variation of incident flux with inclination

1. Single glazing.
2. Double glazing.
E_α: absorbed energy; E_r: reflected energy; E_r: transmitted energy

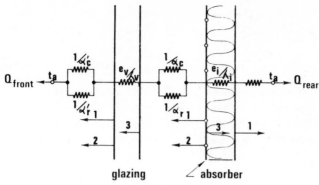

Figure 54. Electrical analogue of losses

The collector is often integrated in the building, which results in a poor value of K_{rear}.

Front Face: The calculation for this is much more complicated and we are simply giving the result.

$$Q_{front} = \frac{T_m - T_a}{\left(\dfrac{c}{T_m}\right)[T_m - T_a/N + f]^{0.33} + \dfrac{1}{h_v}} + \frac{\sigma(T_m^4 - T_a^4)}{\dfrac{1}{\epsilon + 0.05N(1 - \epsilon)} + \dfrac{\epsilon_v}{2N + f - 1} - N}$$

(42)

$f = (1 - 00.4\,h_v + 0.0005\,h_v^2)(1 + 0.091\,N)$
$C = 365.9(1 - 0.00883\,s + 0.0001298\,s^2)$
$\sigma = 5.68 \times 10^{-8}$ W m^{-2} K^4 (Boltzmann's constant)
    $= 4 \times 10^{-8}$ kcal h^{-1} K^4 m^{-2}
s = inclination of collector in degrees (0°: horizontal)
N = number of coverplates
ϵ = emissivity of the absorber
ϵ_v = emissivity of the coverplates
h_v = heat transfer coefficient between the coverplate and outside air

    $h_v = 9 + 3.8\,V$, V = windspeed in m s^{-1}

T_m = average temperature of the absorber (in K)
T_a = average temperature of the ambient air (in K).

The formula can be applied in the following conditions:

$$\left.\begin{array}{l} 320\text{ K} \leqslant T_m \leqslant 420\text{ K} \\ 260\text{ K} \leqslant T_a \leqslant 310\text{ K} \end{array}\right\} \quad \text{i.e.} \quad \left\{\begin{array}{l} 47\,°\text{C} \leqslant T_m \leqslant 147\,°\text{C} \\ -13\,°\text{C} \leqslant T_a \leqslant 37\,°\text{C} \end{array}\right.$$

$$0 \leqslant V \leqslant 10\text{ m s}^{-1}, \quad 1 \leqslant N \leqslant 3, \quad 0° \leqslant s \leqslant 90°.$$

The above expression agrees well with experimental results.

Simplified equation for collector losses:

For practical applications one uses experimental data agreeing with certain types of flat plate collectors. Very simply, a conductance of total losses K is introduced.

$$Q_p = K(t_m - t_a) \tag{43}$$

VALUES OF K

($t_a = 10\,°$C, $V = 5$ m s^{-1}, $\epsilon = 0.95$ non-selective, $\epsilon = 0.10$ selective)

$t_m = 40°$ unglazed non-selective absorber $K = 22$ W m^2 $°$C^{-1}

$t_m = 100°$ single glazed non-selective absorber $K = 8$ W m^2 $°$C^{-1}

$t_m = 100°$ single glazed selective absorber $K = 4$ W m^2 $°$C^{-1}

$t_m = 100°$ double glazed non-selective absorber $K = 4.3$ W m^2 $°$C^{-1}

$t_m = 100°$ double glazed selective absorber $K = 2.5$ W m^2 $°$C^{-1}

$t_m = 100°$ Tedler plastic non-selective absorber $K = 9.2$ W m^2 $°$C^{-1}

$t_m = 100°$ double Tedler non-selective absorber $K = 5.2$ W m^2 $°$C^{-1}

On average one can take

single glazed non-selective absorber $K = 8$ W m^2 $°$C^{-1}

and double glazed non-selective absorber $K = 4$ W m^2 $°$C^{-1}.

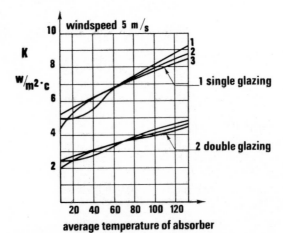

Figure 55. Wind effects on an absorber with emissivity = 0.95.

1. External ambient temperature 40 $°$C.
2. External ambient temperature 10 $°$C.
3. External ambient temperature 20 $°$C.
   (Based on *Duffie*)

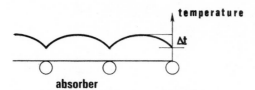

Figure 56. Temperature gradient on a collector surface

Effect of wind on a collector The windspeed appearing in h_v has an influence upon K and as a consequence upon the size of losses. This is why it is important to know meteorological data on wind.

Thermal balance and equations for the collector First equation, or characteristic equation

$$Q_u = G\overline{\tau\alpha} - K(t_m - t_a) \qquad (44)$$

where Q_u represents the energy actually usable.

As we know $G/\overline{\tau\alpha}$ (from meteorological data) and t_a, we can calculate the power Q_u extracted if one determines t_m. This temperature t_m is difficult to estimate in practice because there are temperature gradients on the absorber.

A drop in temperature is produced along the fluid pipes, and conventionally one defines $t_m = (t_e + t_s)/2$ (see paragraph on temperatures reached) and introduces the number F', without dimensions, called the absorber efficiency.

$$F' = \frac{\text{quantity of real heat extracted}}{\text{quantity of heat collected by an isothermal absorber at temperature } t_m}$$

that is, $F' = 1$ for an ideal absorber.

Second equation

$$Q_u = F'\left[G\overline{\tau\alpha} - K\left(\frac{t_e - t_s}{2} - t_a \right) \right] \qquad (45)$$

F' is therefore a controlling parameter for the collector and depends upon the geometry and materials used.

Example of values for F'

Total thickness (in mm)	SPACING BETWEEN PIPES (mm)			
	75	100	125	175
Copper 0.25	0.945	0.970	0.890	0.807
Aluminium 0.50	0.945	0.920	0.885	0.825
Steel 0.50	0.890	0.825	0.750	0.625

$\left. \phantom{\begin{array}{c}a\\a\\a\end{array}} \right\} F'$

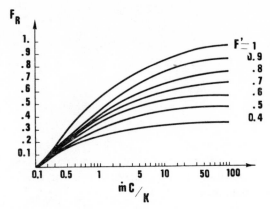

Figure 57. Total efficiency F_R of a standard collector plate

Third equation (more general)

This is an equation for the collector taking into account the flow of the heat transfer fluid. One can observe when making measurements on the performances of a given collector that the entry temperature of the fluid is of great importance. To take this temperature into account, a coefficient F_R is introduced which allows the equation to be written in the form

$$Q_u = F_R [G\overline{\tau\alpha} - K(t_e - t_a)] \tag{46}$$

where

$$F_R = \frac{\text{quantity of real heat extracted}}{\text{quantity of heat collected if absorber is at temperature } t_e}$$

As a consequence it can be shown that F_R is related to the flow of the heat transfer fluid $\dot{m}$ by the expression

$$F_R = \frac{1 - \exp(-F'K/\dot{m}C_p)}{K/\dot{m}C_p} \tag{47}$$

$\dot{m}C_p/K$ is a dimensionless number, C_p is the thermal capacity of the heat transfer fluid.

The collected power as a function of fluid flow can be expressed as

$$Q_u = \dot{m}C_p(t_s - t_e) \tag{48}$$

where Q_u is given by one of the three preceding equations.

Measurement of the output and temperature change for the heat transfer fluid in passing through the collector together with Equation 48 allows us to determine experimentally the amount of collected heat and to calculate the performance of the collector.

It is established that outputs are always low and heat transfer to the absorber fluid is best for a small thickness of fluid.

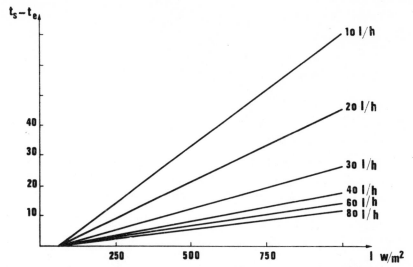

Figure 58. Output of a collector under artificial insolation performed by C.E.A.

Collector performance By analogy with heat engines one can introduce the idea of collector efficiency.

$$\eta = \frac{\text{useful energy collected by the collector}}{\text{solar energy incident upon the collector}}$$

It is important to note that the conditions of use (the sun offers a variable supply) necessitate the definition of efficiency over any finite time interval $\Delta\theta$ as

$$\eta = \frac{\displaystyle\int_{\theta_1}^{\theta_2} \dot{m} C_p (t_s - t_e)\, d\theta}{\displaystyle\int_{\theta_1}^{\theta_2} G\, d\theta} \tag{49}$$

One can therefore measure efficiency instantaneously or on a daily, monthly or annual basis.

In order to compare different collectors one uses the instantaneous efficiency which, using the collector equations, can be written as

$$\eta = \overline{\tau\alpha} - \frac{K_1\left(\dfrac{t_e + t_s}{2} - t_a\right)}{G} \tag{50}$$

assuming the absorber efficiency $F' = 1$.

Examples:

$$\eta_1 = 0.8 - \frac{8\left(\dfrac{t_e + t_s}{2} - t_a\right)}{G} \text{ for a single glazed collector}$$

or $\quad \eta_2 = 0.6 - \dfrac{4\left(\dfrac{t_e + t_s}{2} - t_a\right)}{G}$ for a double glazed collector.

If $G = 1{,}000$ W m^{-2}, $t_a = 15\ ^\circ$C, $t_s = 70\ ^\circ$C, $t_e = 20\ ^\circ$C.

$\eta_1 = 0.8 - 0.24 = 56\%$

and when $t_e = 10\ ^\circ$C,

$\eta_1 = 0.8 - 0.20 = 60\%$.

These two examples show the effect on the efficiency of the mean fluid temperature and show that we have a particular interest in obtaining the lowest possible temperature (t_e) at entry for a given stable exit temperature.

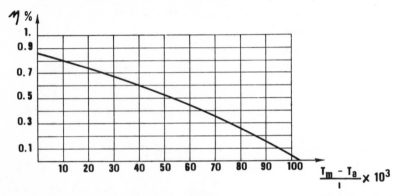

Figure 59. Curve showing efficiency (%) as a function of $(T_m - T_a)/G \times 10^3$, where G is the incident flux

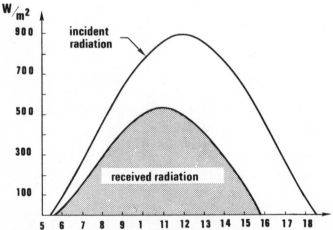

Figure 60. Variations in power output by a collector for a typical day, 21st June (ambient temperature of 15 °C) with a non-selective collector plate, single-glazed, in the South of France

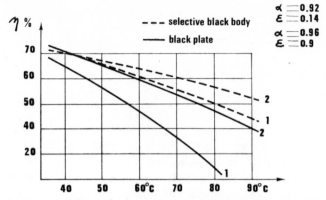

Figure 61. Effect of selective coatings and number of cover-sheets on the efficiency of the collector and temperature level.

1. single glazed. 2. Double glazed

Minimum radiation threshold This is used to determine the available energy on a given day because below this value the collector cannot produce any useful heat.

In effect, since the fluid is circulating the output falls steadily to zero if insolation stops. If one wishes to maintain a constant output temperature, one will need a radiation level given by

$$G_S(t_s) = \frac{K}{\tau\alpha}\left(\frac{t_s + t_e}{2} - t_a\right)$$

(51)

Example: Consider the curves of energy received on two days together with their average value (see Figure 62).

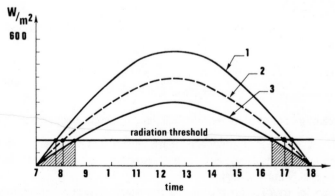

Figure 62. Variation in received energy for two successive days, radiation threshold

1. energy received on the 1st day
2. average energy received for both days
3. energy received on the 2nd day

For a single-glazed non-selective collector

$$I_s(t_s) = \frac{8}{0.8}(30 - 5) = 250 \text{ W m}^{-2},$$

taking $t_e = 10\,^{\circ}\text{C}$, $t_a = 5\,^{\circ}\text{C}$, $t_s = 50\,^{\circ}\text{C}$.

For the *first day*, the radiation threshold is exceeded from 7 o'clock in the morning onwards when the fluid begins to circulate.

On the *second day*, the radiation threshold is not crossed until 9 o'clock. Thus part of the solar energy serves merely to heat the fluid and is therefore not usable, hence the importance of investigating the shadows cast (see Section 2.7). The notion of an average over long periods leads to errors in the estimation of the collected energy and has an important effect on the efficiency.

This concept should be used with caution. The average curve shows that the collector functions from 8.00 to 1600 h, but this is not true for the second day. There will be an input shortfall.

Radiation threshold in a static system   This is used during a summer period to find the null output and to determine the maximum temperature t_{max} achieved with a radiation maximum G_{max} (for durability and reliability of installation).

In effect, once the storage temperature is reached the fluid can continue to heat up without circulating, which could damage the system.

At a certain value of incident radiation G_S the flow is cut off:

$$\dot{m}C_p(t_s - t_e) = G\overline{\tau\alpha} - K(t_m - t_a).$$

The system becomes static; the absorber rises to the uniform temperature $t_s = t_m$ and there is equilibrium between losses and the incident radiation:

$$G_S(t_s) = \frac{K}{\overline{\tau\alpha}}(t_s - t_a) \tag{52}$$

Example: For a fixed output temperature of $60\,^{\circ}\text{C}$ assuming a single glazed collector and $t_a = 10\,^{\circ}\text{C}$,

$$G_S(60\,^{\circ}\text{C}) = \frac{8}{0.80}(60\,^{\circ}\text{C} - 10\,^{\circ}\text{C}) = 500 \text{ W m}^{-2}\,^{\circ}\text{C}^{-1}$$

4.1.12 Connection between meteorological data and collector performances

Each month one can follow the charts giving the number of hours when collected energy exceeds a certain value and correlate this with the efficiency of the collector (see also Section 2.6.9).

This is a useful tool for estimating the necessary storage if the requirements are known (calculated using G and degree-days, monthly or in ten day periods). *Example of calculation:* One can read from the graph (Figure 63) 160 W m^{-2} received energy for 11½ hours with an efficiency of 23%, which gives

$$11.5 \times 160 \times 0.23 = 423 \text{ Wh m}^{-2}.$$

4.1.13 Collector operation under variable conditions

Throughout the day during overcast periods there is an absence of direct radiation, followed by a number of clear periods. These climatic fluctuations cause transitory phenomena in the operation of the collector (the output being stopped if the absence of radiation is prolonged).

It is, therefore, important to know the response times for the collector and for the fluid which transfers the heat to the store. Finally, as we have already seen for the morning, there will be a warm up period and useful operation will not begin immediately.

Studies under variable conditions (analogous, for example, to the charge and discharge of a capacitor in an electrical circuit) illustrate the time constants which influence the response of the system to changes in different parameters; in the case of a solar collector one must expressly include its thermal capacity C and that of the fluid C_p.

The results show that the daily loss is about 5% and these climate-induced variations represent about 0.5% of collected energy.

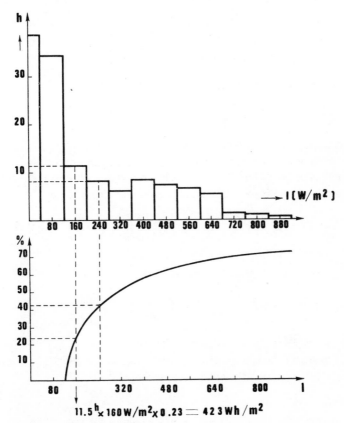

Figure 63. Relation between meteorological data and collector output

Mathematical modelling   The aim is to account for the performance of a solar system over long periods. The significant parameters are:

F_R the efficiency factor for the collector
A the collector surface area
K the heat transfer coefficient for the losses
$\overline{\tau\alpha}$ the optical factor

together with those parameters determined by the application; for domestic applications the production of hot water can be defined by the quantities of heat necessary to ensure usable temperatures.

The equation for a collector with surface area A can be written

$$q_u = AF_R(G_\theta(\theta)\overline{\tau\alpha} - K(t_e - t_a)) \tag{53}$$

where $G_\theta(\theta)$ is the instantaneous value of the radiation on the collector surface.

By using the sinusoidal approximation for the radiation throughout the day

$$G_\theta(\theta) = G_m \cos\frac{\pi\theta}{\theta_d} \tag{54}$$

where G_m is the average mid-day value for a given month, θ_d is the length of the day, taken as constant throughout the month, the thermal balance can be written:

collected energy = energy transferred by the fluid.

The instantaneous equation can be written:

$$\rho C_p \frac{\partial T(\theta)}{\partial \theta} = AF_R(G_\theta(\theta)\overline{\tau\alpha} - K(t_e - t_a)) \tag{55}$$

In winter, temperature varies relatively little and can be considered constant. This, however, is not true in summer where the calculation will be less accurate.

This is a first order differential equation, whose second term is dependant upon time, that is solved by the method of variation of constants to give

$$\frac{t - t_{amb}}{t_s - t_{amb}} = \frac{1 - C_1 G_m \cos\alpha\theta_0}{t_s - t_{amb}} - \frac{C_2 G_m \sin\alpha\theta_0}{t_s - t_{amb}} \exp{-\beta_1(\theta - \theta_s)}$$

$$+ \frac{C_1 G_m \cos\alpha\theta}{t_s - t_{amb}} + \frac{C_2 G_m \sin\alpha\theta}{t_s - t_{amb}}$$

$$C_1 = \frac{(AF_R\theta_d)^2 K\overline{\tau\alpha}}{(\pi\dot{m}C_p)^2 + (AF_RK\theta_d)^2}; \quad C_2 = \frac{\pi AF_R\theta_d\dot{m}C_p\tau\alpha}{(\pi\dot{m}C_p)^2 + (AF_RK\theta_d)^2}; \quad \beta_1 = AF_RK/\dot{m}C_p$$

$$\theta_0 = \tfrac{1}{2} \text{arc} \cos(K(t_s - t_{amb})/G_m\overline{\tau\alpha})$$

where t_s is the storage temperature at the beginning of the day. t_{max} for the end of the day is obtained by using the above expression, and

$$t_s \geqslant t_{min} \text{ where } t_{min} = 37\,°C.$$

In this way it is possible to predict the variation of storage temperature from meteorological data (variable radiation and temperature).

For practical applications, allowing for the uncertainties in the meteorological data, the given equations are adequate.

4.2 SEALED WATER COLLECTOR

The circulation of the fluid can either be natural (thermosiphon) or forced by a circulation pump.

Description of collector. This is a box made of galvanized sheet with a covering which is transparent to the incident radiation. Differences between collectors arise from the design of the absorber and the coverplates (single or double glazing, type of glass, selective, or non-selective absorber, design of heat exchange surfaces).

There are three types of absorber:

4.2.1 With a serpentine pipe

The pipe is generally copper, soldered to a thin copper plate covered with an absorbing paint. This collector was common in the U.S.A., South Africa and Spain in the 1960's.

4.2.2 With parallel pipes and channels with manifolds

Drawn round or oval pipes are connected to two manifolds. The pipes can be horizontal or inclined. Oval pipes with their side of largest curvature exposed to the sun have the advantage of containing less water for the same surface area exposed to incident radiation and will provide hot water shortly after sunrise.

The starting time can be further reduced by using rectangular conduit containing only a narrow layer of water.

This process is used in Israel and the U.S.S.R. (Tashkent). A recent development consists of placing the fluid veins directly into a metallic plate (tube-in-strip or Roll-band with aluminium). This is less expensive than tubes fastened to the plate and the contact is better; consequently heat transfer between the plate and the pipe is better.

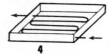

4

Figure 64. Collector plate with serpentine pipe Type 4.

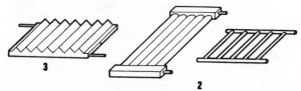

3

2

Figure 65. Collector plates with manifolds Types 2 and 3.

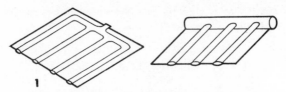

Figure 66. Tube and strip absorbers Type 1.

The collectors are frequently inclined at an angle to the horizontal plane. The water at the bottom is cooler than at the top.

Since the temperature difference between the water and the absorber-piping assembly is larger at entry, the lower part of the absorber is more effective than the top.

4.2.3 With plates

In its most simple form this consists of two metal plates soldered along the edges. For pressurized applications it is necessary to use extra rivets to reinforce the assembly.

Parameters

distance between the plates: 10 mm
water capacity: 1 to 15 $1 \, m^{-2}$
total weight of collector: 60 kg

Type	Useful Area	Total Area	Length m	Width m	Height mm
1	3 m^2	3.42 m^2	3.105	1.120	130
2	2 m^2	2.32 m^2	2.074	1.120	130
3	1.6 m^2	1.93 m^2	2.074	0.93	130

4.3 AIR COLLECTOR

4.3.1 The design of this collector is different for three reasons.

The volume of air flowing is much larger because the thermal capacity of 1 m^3 of air is 0.34 Wh $°C^{-1}$, whereas that of 1 m^3 of water is 1,160 Wh $°C^{-1}$.

Under normal conditions, the warm air produced by a double glazed solar collector is 30 to 50 °C above the temperature of the input air; in a water collector the level of exit temperature reaches 35 to 60 °C. This difference is because the convection heat transfer coefficient from a surface to air is much smaller than that to a liquid.

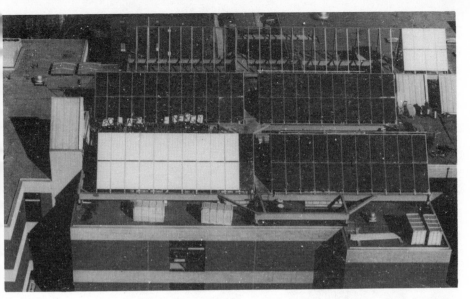

A building at Dorchester, U.S.A. The collector array is on the roof.

(Photo: ISES)

An industrial building at Phoenix, Arizona, U.S.A. Some of the collector surfaces overhang and act as sunshields.

(Photo: ISES)

Private dwelling in Colorado, U.S.A.

(Photo: Nicolas and Vaye)

An industrial building in Germany showing the collectors being mounted.

(Photo: the author)

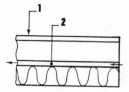

Figure 67. Cross-section of a collector using air.
(Length − 3.65 m; width − 59.6 cm; thickness
of air passage − 1.49 cm)

air in free convection 5.8 to 29 W m^{-2} °C^{-1}
oil 58 to 11,740 W m^{-2} °C^{-1}
water under forced convection 290 to 11,600 W m^{-2} °C^{-1}

The third difference arises through the type of storage and the patterns of use associated with each type of collector; this point will be discussed further in paragraph 4.5 (Storage).

4.3.2 Test results

It has been possible to compare the output of a collector using air with that of a collector using water under identical operating conditions.
The trials were performed under the following conditions:
(1) 0.763 m^3 min^{-1} m^{-2} corresponding to an airspeed of 3.67 m s^{-1},
(2) 0.431 m^3 min^{-1} m^{-2} corresponding to an airspeed of 2.07 m s^{-1}.

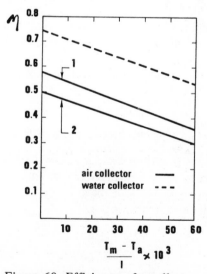

Figure 68. Efficiency of a collector
using air compared with efficiency
of a collector using water.
Flows (1) = 0.763 m^3 min^{-1} m^{-2}
(2) = 0.431 m^3 min^{-1} m^{-2}
in a collector of 32 m^2 (14
collectors in series)

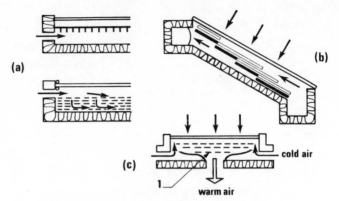

Figure 69. (a) collector with finned metal plate.
(b) glass sheets of which $\frac{1}{3}$ are blackened.
(c) Bliss collector (1). Partially blackened glazing

The efficiency curves are expressed as a function of

$$\frac{\frac{t_e + t_s}{2} - t_a}{I} \left(\frac{m^2\,{}^\circ C}{W}\right)$$

Examples of hot air collectors

(1) collector with a finned metal plate — the fins are usually at right angles.

(2) collector using the greenhouse effect and heat conduction: the collector has multiple functions — a thermal function (winter heating, summer ventilation) and an architectural function in that the conducting wall forms the envelope of the building.

4.4 TRICKLE-FLOW COLLECTOR (FREE FLOW UNDER GRAVITY)

The principle is to warm a fluid while it is flowing slowly from the top to the bottom of the collector. It is fundamentally different from the sealed type of collector where water circulates from the bottom to the top due to the thermo-siphon effect of the temperature difference.

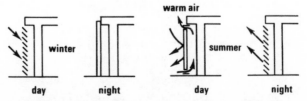

Figure 70. Collector with variable insulation

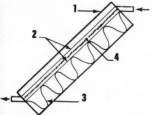

Figure 71. Water flow in a porous structure.

1. Glazing.
2. Metal.
3. Insulation.
4. Porous material

4.4.1 Use of a porous structure

The water circulates through a porous structure between two metal plates in a box fitted with a transparent coverplate. The detrimental effects of freezing are avoided.

4.4.2 A collector having an architectural function

A collector can be employed as an element of roofing in the form of corrugated metal, ribbed asbestos sheet or an aluminium trough. The heat transfer fluid can be oil (whose viscosity is much larger than water) or air, but there is a risk of corrosion because of condensation.

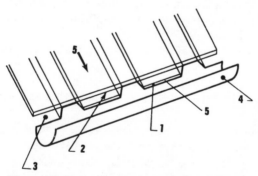

Figure 72. Example of a trickle-flow collector

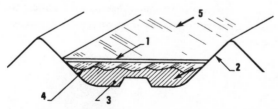

Figure 73. Collector using oil flow.

1. Glazing.
2. Trough-shaped roof.
3. Oil flow.
4. Blackened sheet metal.
5. Direction of flow.

The fluid is collected in a gutter at the bottom, from where it flows to the store. It is then pumped back to the top of the collector.

There is an inherent economic advantage in this system as materials and installation costs are low making these uncomplicated collectors fairly inexpensive.

4.5 CONCENTRATING COLLECTOR SYSTEMS

Concentration is obtained by reflection of the solar radiation from plain or curved mirrors. The reflected radiation is concentrated in the focal zone, thus increasing the energy received per unit of surface area in this zone.

The concentration factor depends upon the apparent diameter of the sun, which varies seasonally, upon the orientation of the system to the sun and upon the type and condition of the surface (staining, oxidation).

4.5.1 Advantages of concentration

(a) It reduces the absorber area required and consequently reduces the thermal losses (proportional to collector area). These losses also depend upon the difference in temperature between the absorber and external air but as the exchange surfaces are smaller the advantage is maintained.

(b) For identical losses a concentration system permits a much higher temperature to be reached. This is important for some applications, such as solar motors, water pumping and steam production.

(c) For the same flux the use of reflectors allows the absorber temperature to reach a higher value.

Pure aluminium (with a reflective power almost as high as for silver) is excellent in the ultraviolet and in the infrared. It is inexpensive and one can use sheets of polished or anodized aluminium.

The most popular reflectors are curved because of their simple fabrication. By contrast, a flat reflector must have sufficient rigidity to preserve its flatness, and a mirror of thin sheet metal can only be flat if it is under tension. The amount of material used for a curved mirror is smaller than would be needed for a flat collector plate (see Figure 74).

The much higher temperature of the concentrating collector allows the storage of more heat in the same volume. The cost of storage is, therefore, less than in non-concentrating system.

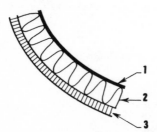

Figure 74. Schematic cross-section of a concentrating mirror

For solar refrigeration and air-conditioning systems it is necessary to achieve temperatures of at least 90 °C in order to obtain reasonable efficiencies, which means double glazing and a selective absorber are necessary when a flat plate collector is used.

However, in a concentrating system the smaller absorber area needed for the same duty makes possible both the use of selective surfaces and a vacuum between the absorber and the glazing.

The amount of antifreeze liquid is smaller than in a water-circulation collector.

4.5.2 Drawbacks of concentration

Concentration systems focus the direct radiation onto the absorber. They collect less of the diffuse radiation than flat plate collectors, and they are not suitable for use in very cloudy regions.

4.5.3 Orientation with respect to the sun

Exact orientation involves tracking in two parameters: the altitude and the azimuth. This tracking is essential for solar furnaces. If we consider only the altitude, two approaches are possible: a fixed absorber and a mobile concentrator, or *vice versa*.

The present trend in the U.S.A. is to choose the second solution for building applications on account of its flexibility of use and better economy.

A photo-electric cell is used to track the sun and this determines the movement of the absorber.

The reflection factor of the mirror deteriorates with time due to staining or pitting of the aluminium etc.

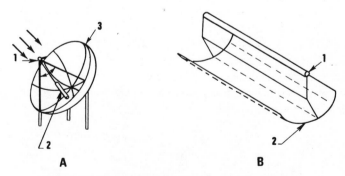

Figure 75. Examples of a simple tracking system.

    (A) 1. Photocell.
          2. Mobile absorber.
          3. Fixed spherical mirror.
    (B) 1. Mobile absorber.
          2. Parabolic cylinder.

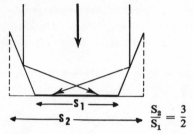

$$\frac{S_2}{S_1} = \frac{3}{2}$$

Figure 76. A concentrator with flat mirrors

4.5.4 Concentration factor

This is the relation between the normal surface exposed to the sun and the image of the sun formed on the absorber by the system.

Example:

(1) A concentrator with flat mirrors (Figure 76)

$$C_r = \frac{3}{2} = 1.5$$

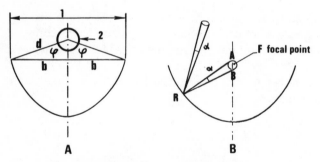

Figure 77. Cylindrical parabolic concentrator.
1. Receiving surface, 2. absorber

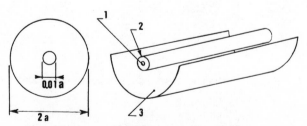

Figure 78. Parabolic reflector.
1. Tube, 2. casing, 3. reflecting surface

(2) Cylindrical parabolic mirror

The divergence of reflected rays is equal to the angle α subtended by the sun at the earth (the apparent diameter of the sun), and is $32'$ of arc (0.00931 radians).

The axis of the cylinder lies in an east-west direction and this collector is used for solar water heating. The focal line is formed by a tube filled with water. The image of the sun has a diameter $AB = 0.00931 \times d$, with $d = RF$ (in the triangle RFA, $AF = tg(\alpha/2)d = 0.00465d$ and $AB = 2\ AF$).

The parabolic cylinder is characterized by the position of the focal zone and by its aperture.

$$C_r = \frac{\text{collecting surface}}{\text{absorber surface}} = \frac{2 \sin \phi \times d \times \text{length}}{0.0093 \times d \times \text{length}}.$$

If $\phi = 90°$

$$C_r = \frac{2d}{0.0093d} = 200.$$

Example of attained temperature: the incident energy G is reflected, concentrated and absorbed; the energy gained is $Q_u = GRC_r\alpha$.

Thermal equilibrium is reached when the heat lost is equal to the heat gained.

The energy lost is $Q = \sigma \epsilon T^4$

$$\therefore \qquad GRC_r\alpha = \sigma \epsilon T^4$$

where ϵ is the emissivity of the absorber and $\sigma = 5.77 \times 10^{-8}$.

Numerical example: if $G = 600\ \text{W m}^{-2}$, $R = 0.9$, $\alpha = 0.95$, $C_r = 5$, $\epsilon = 0.9$, then $T = 198\,°C$.

Concave spherical reflector of radius R: The image is formed at a point on the axis distance $R/2$ from the reflector surface.

$$C_r = \frac{\pi R^2}{(0.0093R)^2 \pi} \simeq 100$$

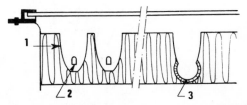

Figure 79. Parabolic concentration in a flat plate collector.

1. Parabolic concentrator formed from two parabolic sections symmetrically placed with respect to the absorber,
2. the absorber tubes,
3. alternative arrangement

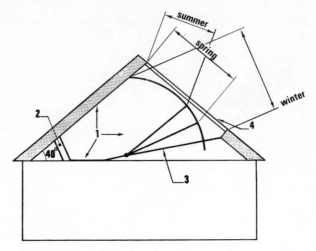

Figure 80. Pyramidal mirror system with air or water as fluid; advantages – protection against inclement weather and ice, an optical gain after reflection of 1.6 to 4.8 compared to a flat plate collector.

1. Reflecting surface, 2. flat plate collector, 3. mobile reflector, 4. window

Parabolic reflector: This is obtained by the rotation of a parabola about its axis:

$$C_r = \frac{\text{collecting surface}}{\text{absorber surface}} = \frac{(2a)^2\,\pi/4}{(0.01a)^2\,\pi/4} = 200^2$$

therefore, the theoretical concentration factor C_r is 40,000.

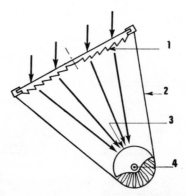

Figure 81. Fresnel lens system; axis of orientation East–West but tilted according to the season.

1. Fresnel lens, 2. support structure, 3. transparent window, 4. absorber tube

4.6 STORAGE OF SOLAR ENERGY

One of the desired aims is to assure the supply of all demand for a certain period. However, it is not economically possible in our temperate climate to provide whole-year storage with the techniques currently available.

The simplest technique uses the sensible heat of matter. The heat transfer fluid circulating in the collector absorbs a quantity of heat and then heats another fluid contained in a storage vessel. For example, consider a house with a heating demand of 15,000 kWh per year. The collector system extracts 7,500 kWh, leaving a shortfall of 7,500 kWh which has to be collected and stored during the summer period.

If water is the storage fluid the thermal capacity is 1.16 Wh kg^{-1} °C^{-1}, and assuming its temperature to be 50 °C one can store 58 Wh kg^{-1} of water. Hence the volume of the store is

$$\frac{7,500}{58 \times 10^{-3}} = 129 \text{ m}^3 \ (107 \text{ m}^2 \times 1.20 \text{ m high}).$$

In winter the length of the day is much shorter (8 h 15' on 21st December), thus reducing the amount of energy received. In addition, there are breaks in the sunshine of varying duration, and these breaks have to be allowed for and should be assessed from the meteorological data. Winter demand is, to a greater or lesser extent, continuous over the 24 hour day.

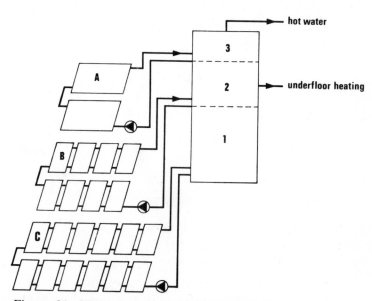

Figure 82. Using three temperature stratifications for different purposes.
A. Double glazed collector with selective absorber.
B. Double glazed collector.
C. Single glazed collector

Under these conditions storage should be designed to supply the demand ov short periods (two nights and one day, or one night), with an indispensable peric of auxiliary heating.

On the other hand, as the size of the store should not be overestimated for co reasons, it is wiser to design around its use during the interseasons (autumn ar spring) when the temperature is not too low and when sufficient solar gains a available.

In hot water stores the liquid stratifies into different temperature layers. Th stratification can be dispersed by introducing turbulence between the layers encourage thermal mixing. This can be supplied by using vertical plates (as in sept tanks) or by a circulation pump, but is only advised for large scale production domestic hot water, say for a block of flats. Stratification is often retained in ord to cope with different requirements (hot water 60 °C, underfloor heating at 40 °C and to provide the lowest possible return temperature.

Examples: Consider a house sited at Tours, in zone B, with a volume of 250 n (100 m² floor area) and well insulated ($G = 1$ W m^{-3} °C^{-1}).

(1) Let us determine the storage volume V needed to supply demand for tw consecutive sunless days in winter (December). The average daily degree-days f the month of December = 15.

heat losses: 90 kWh per day,

thus, for 2 days: 180 kWh.

Assuming an average temperature of 35 °C (the collector supplies water at 30 ° minimum to 50 °C maximum), the heat stored by 1 m³ of water is 1.16 x 35 40.6 kWh;

therefore the storage volume V is $\dfrac{180}{40.6} \simeq 4.4 \text{ m}^3$.

(2) In the inter-season (March)

Average degree-days = 11

heat losses: 132 kWh.

(Heat stored at an average temperature of 45 °C (higher than in winter)):

1.16 x 45 = 52.2 kWh.

$$V = \frac{132}{52.2} = 2.52 \text{ or } 2.5 \text{ m}^3.$$

The storage volume is, therefore, almost doubled in December. It remains t determine the cost of supplementary heating against the savings realized. General the volume of storage is related to collector area so that

100 1 m^{-2} < V < 150 1 m^{-2}.

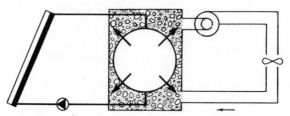

Figure 83. Diagram of storage using stones

4.6.1 Different possibilities for storage

(1) water,
(2) crushed rocks or gravel with a good thermal capacity (e.g. bauxite),
(3) dissolved salts,
(4) heat-induced reactions in mineral oxides.

(1) *Water* is the most often used medium for the following reasons: it is virtually cost-free, although at high temperatures it can only be used for storage in liquid form in a pressurized vessel which raises costs. At low temperatures, it can be stored in galvanized tanks, fibreglass or water-tight concrete vessels, which provide insulation and can be sited in an accessible place. Due to a higher thermal capacity less volume is required than for rock storage.

(2) *Crushed rocks and gravel.* These are used in hot air systems and in air-conditioning. It is easier to store gravel than water as the problem of leakage does not occur. A heat exchanger is not required between the collector fluid and the storage fluid as the rock is its own heat exchanger.

Again taking the preceding example for the inter-season, where $V = 2.5$ m^2 for water at an average temperature of 45 °C, there are voids in a crushed rock storage vessel which account for about 30% of the volume. Therefore, since the stones occupy only 70% of the volume the effective density is 0.7 times that of the stone. Thus, the density of the stone is $2,400 < d < 2,900$ kg m^{-3}, and the specific heat is $C = 0.30$ Wh kg^{-1} °C^{-1}.

Therefore, the heat stored per m^3 is ($d = 2,800$ kg m^{-3}),

$$2.8 \times 0.7 \times 0.30 \times 45 = 26.5 \text{ kWh} \quad \text{and} \quad V = \frac{132}{26.5} = 5 \text{ m}^3$$

(approximately double that of a water store).

(3) *Dissolved salts.* The heat of reaction of certain substances in the temperature region 35° to 70° is used.

One can use eutectic salts (hydrated mineral salts) and sulphate of sodium (Glauber salt).

The salt dissolves by forming an anhydrous salt solution, which releases a large amount of heat to the surroundings as the process is endothermic.

An equal quantity of heat is recovered when the solution is cooled. Crystals of hydrated salt in suspension are formed when the substances are again combined with water.

	Change of state temperature ($^\circ$C)	Heat of reaction (Wh kg^{-1})
$CaCl_2-18\ H_2O$ calcium chloride	29–39	48
$Na_2S_2O_3-5\ H_2O$ sodium carbonate	32–36	74
$Na_2S_2O_3-H_2O$	49–51	50

The amount of heat of reaction compared to the thermal capacity of the water and stone shows the benefit of this approach (smaller storage volume).

Several problems related to this method still have to be solved (corrosion and crystallization of the salts).

(4) *Heat of Reaction of mineral oxides*. The heat of hydration of mineral oxides such as MgO and CaO can be used. The heat of the medium is absorbed, decomposing $Mg(OH)_2$ and $Ca(OH)_2$ into MgO and CaO. The heat is restored by mixing the stored oxides with water in a tank (rehydration).

Advantage: the energy density is much higher than that of the dissolved salts (10 to 30 times higher), and the oxides can be stored at the ambient temperature for long periods (long duration or annual storage).

The decomposition of $Mg(OH)_2$ takes place around 373 $^\circ$C and the decomposition of $Ca(OH)_2$ occurs at about 520 $^\circ$C; therefore, the practical application of this process requires concentration systems which can produce this temperature level.

4.7 DISTRIBUTION AND CONTROL OF THE THERMAL CIRCUIT

The circuit supplied by the collectors has four functions:

(1) *distribution of the extracted heat to a particular location*;

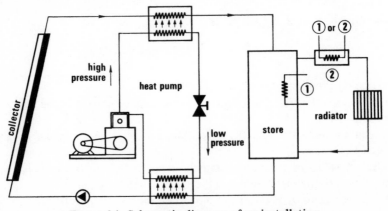

Figure 84. Schematic diagram of an installation

(2) *control of the heat extracted*: demand does not coincide with the solar gains (washing requirements in the morning);

(3) *control of temperature* if the solar radiation is insufficient to allow distribution of the heat at a temperature appropriate to the requirement;

(4) *collector installation* under optimal conditions of operation allowing good circulation of the heat-transfer liquid: the entry temperature of the fluid should be as low as possible in order to reduce losses (achieved, for example, by extraction of heat from the return circuit to the exit circuit).

4.7.1 Distribution

There are generally two circuits: the primary or collector circuit and the secondary or distribution circuit.

In Section 5.2 we will consider the problems of central heating, but it is worth noting here that the temperatures reached in flat plate water collectors (70 ° maximum) do not meet the requirements of traditional central heating (t_s = 95 °C) with a conventional supplementary heat source.

On the other hand the temperatures produced do correspond with low temperature underfloor or blown air heating.

4.7.2 Control of heat

This is controlled by means of the storage selected

4.7.3 Stratification

Depending on the level of insolation the exit temperature from the collector can be lower than that needed for the operation of the system. In this way fluids at different temperatures are obtained which can either be stratified or mixed.

4.7.4 Use of a heat pump to reduce return temperature

A low temperature in the return circuit can be obtained by using a heat pump between the flow and the return lines. This application of the heat pump has been used at Creil where 4,000 apartments are heated geothermally.

An exchanger can be used to heat the new air or to heat a glasshouse which would act as a buffer zone on one façade. It is also worth examining the possibility

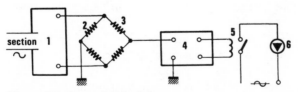

Figure 85. Control principle.

1. Generator, 2. collector sensors, 3. store sensor, 4. differential temperature sensor, 5. mains relay, 6. circulation pump

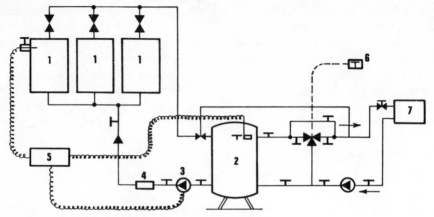

Figure 86. Diagram of the control and installation.
1. Collectors,   2. store,   3. pump,   4. non-return   valve,   5. control   box,
6. thermostat controlling a 3-way valve, 7. radiator

of a two-tank system, one hot tank and one cold tank, so that the cold tank could extract the heat from the return pipe (see the applications in Chapter 5).

The other controls correspond to a traditional heating system.

Control of flow can be used to start or stop the collector at the radiation threshold.

4.7.5 Regulation of exit temperature from the collector to a fixed value which corresponds to the efficient operation of the system.

An array of three-way valves and sensors to calculate the solar heat allows the practical realization of different circuits.

It is also important to note that the control system must prevent operation of the system when the energy extracted from the collectors is less than or equal to the energy used by the control system itself.

4.8 DETERMINATION OF COLLECTOR AREA

The calculations relate to a house of 100 m² floor area, corresponding to a heating volume of 250 m³.

Two examples are given in the table of results:
    Collectors installed on a southern façade, and
    Collectors inclined at 60° on the roof.

The meteorological data used apply to the Carpentras region for an average year, and the calculation of monthly losses is made using the results from Chapter 2.

Solar house at Aramon, Gard, France

(Photo: the author)

Solar house at Aramon

(Photo: the author)

Demonstration solar house at Washington Exhibition, September, 1976, using a concentrating flat plate collector, supplying an absorption system.

(Photo: the author)

Solar house with central heating by water flat plate collectors on the roof (Washington, 1976)

(Photo: the author)

Solar water heater installed at Mejannes–le–Clap, Ardèche, France. Note the array of collectors used to heat water for the swimming-pool

(Photo: the author)

Table I

Inputs (W m^{-2})

Month	(1) G_{OH}	(2) D_{OH}	(3) h	(4) S_{OH}	(5) S_{OV}	(6) $S_{60°}$	(7) D_{OV}	(8) $D_{60°}$	G_{OV}	$G_{60°}$
October	615	154	36.5°	461	623	770	154	162	777	932
November	421	105	29.7°	316	570	611	105	110	675	721
December	369	92	23°	277	653	704	92	97	699	801
January	435	109	24.5°	326	723	764	109	116	832	878
February	577	144	35°	433	618	752	144	178	762	930
March	721	180	45°	541	541	739	180	189	721	928
April	837	209	52.5°	628	481	731	209	219	690	950
May	865	216	64.5°	649	310	592	216	227	526	819

(1) G_{OH} total radiation received at local noon for one day under clear sky conditions on a horizontal plane
(2) D_{OH} = 0.25 G_{OH} diffuse radiation component
(3) h = the altitude of the sun at local noon

$h = 90° - \phi + \delta \begin{cases} \phi = 44° \\ \delta = \text{solar declination} \end{cases}$

(4) S_{OH} direct radiation component $S_{OH} = G_{OH} - D_{OH}$

(5) $S_{OV} = S_{OH} \times \dfrac{\cos u}{\sin h}$, $\cos u = \cos h \cos a$

(6) $S_{60°} = S_{OH} \times \dfrac{\cos u'}{\sin h}$, $\cos u' = \cos a \sin 60° + \sin h \cos 60°$

(7) $D_{OV} = 0.5 (D_{OH} + 0.3 \, G_{OH})$
(8) $D_{60°} = 0.75 \, D_{OH} + 0.075 \, G_{OH}$

93

Table II

					Collector Area				(15) Collector area in m²	
	(9)	(10)		(11)	(12)	(13)	(14)			
Month	G_V	$G_{60°}$	ΔT	Q per day	$Q_{60°}$ per day	D	Losses kWh		vertical/	@ 60°
October	623	738	11	4.371	5.170	60	0.66	360	5.3	4.5
November	479	500	10	2.897	3.180	189	0.52	1,134	32	29.7
December	524	590	9	3.004	3.376	339	0.58	2,034	44	48
January	591	628	9.5	3.574	3.793	303	0.55	1,818	33	38
*February	559	665	10.5	3.561	4.440	237	0.55	1,422	28.5	23
March	557	702	11.5	4.078	5.130	215	0.61	1,290	20.4	16
April	575	779	12	4.395	5.954	106	0.70	636	9.6	7
May	421	643	13.5	3.757	5.52	49	0.65	294	5	3.5

*Reference month for choice of collector area

(9) $G_V = G_{0V}(0.33 + 0.7\sigma)$ (W m^{-2})

(10) $G_{60°} = G_{0.60°}(0.33 + 0.7\sigma)$, ΔT = length of day in hours

σ = insolation fraction

(11) $Q_V = \dfrac{2}{\pi} G_V \Delta T$ energy received per square metre per day

(12) Ditto

(13) D = degree-days at Carpentras $t_i - t_e = 15\,°C$

(14) Losses = $24\,h \times G \times V \times D$ when $G = 1\,\text{W m}^{-3}\,°C^{-1}$

$V = 250\,\text{m}^3$

(15) $24\,h \times G \times V \times D = Q_V \times$ area of collector $\times \eta \times$ days, where

η = efficiency of collector

= 40% (November, December, January)

= 50% (other months).

5

Applications

5.1 THE PRODUCTION OF DOMESTIC HOT WATER

This is the most widespread application of solar energy. The range of temperatures obtained by flat plate collectors corresponds to the temperature of hot water used in dwellings. This technique has been developed over a long time in some countries where there is a high insolation.

Japan has the largest numbers of installations, with over two and a half million; Israel has over 150,000 solar water heaters and more than 25,000 installations are in use in the State of Florida.

The conditions of use differ greatly from one country to the next, and depend upon four variables.

5.1.1 Hot water demand and type of use

This variable is related to sociological criteria (social development, customs, availability of water). For instance, in Japan the regular daily bath can account for the widespread use of these installations. In the U.S.A. the growth in the number of private swimming pools and the development of solar techniques go hand in hand.

The size of an installation is determined by estimating the daily consumption (60 litres per day per person in France and 75 litres per day in the U.S.A.). Consumption increases with social level (two bathrooms in the same apartment).

Storage volume is based on the estimated consumption in order to provide for the demand over a 24 hour period. However, supplementary heating of adequate capacity is necessary at our latitudes (electricity, gas, oil).

Architectural integration poses three problems:
the siting of the storage tank (loft or underground),
the weight of the tank has to be considered when calculating the load on the joists, and
the position of the collectors; this is an important but primarily aesthetic consideration. Water heaters in Israel are installed on roof-tops, giving an overall irregular appearance which is unsatisfactory.

95

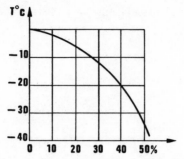

Figure 87. Variation of freezing point of water with percentage of ethelyne glycol added

Scaling can occur in the same way as in a traditional heating system, and the risk of his happening increases at temperatures above 70 °C. However, this risk can be reduced by using an indirect heat circuit and a heat exchanger in the tank.

Winter temperatures necessitate anti-freeze protection, and this is obtained by mixing water and alcohol or water and an ethylglycol mixture in appropriate proportions.

There are drawbacks to using glycol (oxidation, instability of the solution) which justify not using this anti-freeze in solar water heaters. Some suppliers provide a detector which sets a pump in motion when the temperature drops to 0 °C in the collector; or, to allow for periods of deep frost, a butane heater controlled in the same manner can be installed at the base of the collector.

In the majority of cases the size of solar water heaters in general use corresponds to family demands. Large sized water heaters are only feasible for apartment buildings, public baths, schools, hotels, or hospitals.

5.1.2 Scale of usable solar energy

The meteorological data (solar radiation intensity G, sunshine hours, the number of days without sun and the number of days with sun), air temperature and the temperature of water admitted to the equipment are obviously the determining conditions. Every region with more than 1,800 hours of sunshine per year can be considered a suitable installation zone.

5.1.3 Orientation and inclination of collectors

Hot water demand is almost constant throughout the year and we have already seen that a fixed inclination for the collector (the latitude of the site) can be used for year-round use; an even better approach would be to use two settings, one for winter (ϕ +12°) and one for summer (ϕ −12°). If ϕ +23.5° is selected the collector will be pointing towards the position of the sun at its winter zenith.

5.1.4 Different types of solar water heaters

There are two main categories:

(1) heat absorption and storage in the same assembly, and
(2) heat absorption with separate storage.

The first type has been developed in several forms, particularly by the Japanese, usually in a plastic material and at a very low price, comparable to the price of an electric storage heater.

NOTE: Japan (except for Hokkaido) lies between 30° and 40° north latitude. It has a moderate climate and a high solar radiation intensity.

Examples of Japanese water heaters

The second type of heater keeps the two functions of absorption and storage separate as is shown in Figure 91.

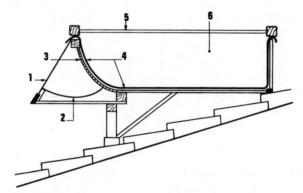

Figure 88. Japanese water heater type 1.

1. Glazing, 2. reflector, 3. wire mesh, 4. black vinyl,
5. glazing, 6. water

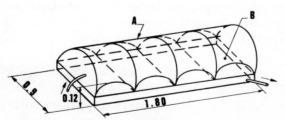

Figure 89. Japanese water heater type 2, plastic bag containing 200 litres connected in series.

A. Transparent vinyl covering,
B. Black vinyl tank

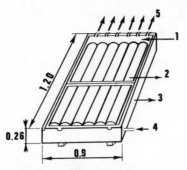

Figure 90. Japanese water heater type 3, closed box with a cover-plate and 6 black plastic cylinders, 80 litres capacity

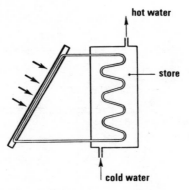

Figure 91. Schematic diagram of solar water heater

The water is heated in the collector and begins to circulate through natural thermal circulation or by using a pump.

The various models in this category differ in the absorption surface chosen (see the different sketches of absorbers in Section 4.1.2), in the control of the system and in the stratification into more or less independent zones in the tank.

(1) *The natural circulation of water: thermosiphon.* This can only take place if the storage tank is sited at a certain height above the collector.

This is the simplest system and the most extensively used, especially in Israel and Australia.

There are, however, two drawbacks:

(a) the storage must be close to the collectors, which imposes constraints on the ᵗ and weight of the tank;

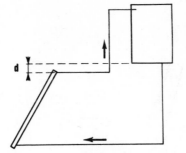

Figure 92. Natural circulation of water

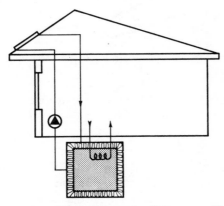

Figure 93. Schematic diagram with underground store

(b) when the water is circulating by thermosiphon the installation is self-regulated. One cannot, therefore, influence the output which depends upon the value of the solar radiation (see the curve $Q_u = \dot{m}C_p(t_s - t_e)$ for different values of the incident radiation).

(2) *Circulation pump and control.* A low power circulation pump in the return circuit removes the restrictions on the forced position of the storage in relation to the collectors. Another advantage is that a controlled variable flow can be obtained in such a way that it is possible to get the same increase in temperature even when the received radiation varies.

Example: in order to maintain a ΔT of 30 °C with an incident radiation of

$$400 \text{ W m}^{-2} \quad \dot{m}C_p = 8 \text{ l h}^{-1} \text{ m}^{-2}$$
$$300 \text{ W m}^{-2} \quad \quad \quad 5 \text{ l h}^{-1} \text{ m}^{-2}$$
$$200 \text{ W m}^{-2} \quad \quad \quad 3 \text{ l h}^{-1} \text{ m}^{-2}$$

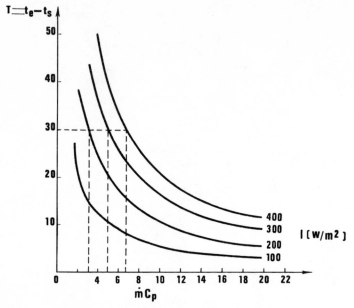

Figure 94. Expression for the temperature rise

The other advantage of control circuitry is that it can switch off the pump to prevent water circulation through the collectors until, for example, $T_m = T_S + 5\,^\circ\text{C}$, where T_m = average temperature for the collectors.

5.1.5 Examples of water heaters manufactured in France

(1) SOFEE system

Stratification in the storage is ensured by plates which prevent movement between layers of fluid at different temperatures.

An electrical thermostat in the upper part of the tank monitors the temperature and, if required, switches on electrical supplementary heating of the water to ensure that it is at the temperature necessary for use. Finally, the return flow travels through the bottom of the storage, where the layers of water are coldest, in order to obtain the best possible efficiency.

(2) JOHANES system

This system is based on two panels and two tanks.

We have seen that the efficiency increases as the temperature of the return from the store decreases (Figure 59, collector efficiency curve). A way of achieving this using a heat pump is shown in Figure 84.

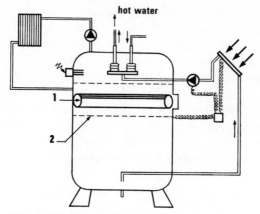

Figure 95. SOFEE Water-heater.

1. Expansion vessel, 2. perforated plates to keep temperature stratification

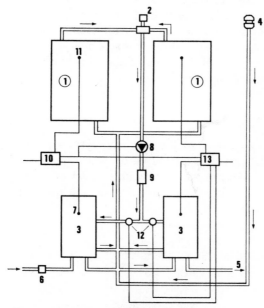

Figure 96. JOHANES Water-heater

1. Collector, 2. vent, 3. stores, 4. expansion vessel, 5. hot water, 6. safety valve, 7. sensor for differential thermostat, 8. pump, 9. non-return valve, 10. electronic amplifier, 11. sensor on the solar panel, 12. solenoid valve, 13. electronic amplifier for differential and maximum control

Another solution uses two storage vessels which are linked by a control system. Once operating temperature is reached in the first tank a valve closes the circuit and opens access to the second tank which can, therefore increase its thermal capacity quickly as it is at a low temperature (cold water supply).

In each case supplementary heating will supply water at the service temperature (90 °C for central heating). This can be achieved either by additional resistance heaters in the storage vessel or by installing a supplementary heater (e.g. a gas water-heater) on the distribution pipe. When this approach is used the storage tank supplies the water at its own temperature (50 °C), acting as a preheater. The storage capacity corresponds to approximately 120 litres per square metre of collector.

5.2 THE HEATING OF BUILDINGS

Systems using solar gains fall into the three following main divisions:

5.2.1 Passive systems

The shell of a building acts as a collector (solid surfaces and glazed surfaces). Insulation on the outside of the building allows the solar radiation to be absorbed by the internal walls after entering through the glazed openings (east, south, west).

These glazed openings should be fitted with external shutters to prevent nocturnal heat losses. One current process consists of injecting small balls of polystyrene between the panes of double glazing, which increases the insulation and limits the negative greenhouse effect.

Different approaches

(a) *A heavy masonry wall:* All the functions of the system (collection, storage, transfer, and control) are concentrated on the vertical southern façade. This wall, which is constructed of heavy material, is painted black on the outside.

A rigid framework supporting a single or double glazing is erected at about 12 cm in front of the wall.

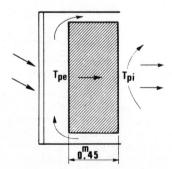

Figure 97. Masonry wall

Ventilation openings of section 0.84 × 0.095 m are arranged at the top and bottom so that air can circulate by natural convection.

The heat collected in the wall passes by conduction towards the inner face where a heat exchange is produced by radiation and convection. This quantity of heat is available after a time lapse which depends upon the nature and thickness of the material used (see Section 3.3).

For a light concrete, $\rho = 800$ kg m^{-3}, $e = 15$ cm and the time is 5 hours.

When using concrete of density $\rho = 2,450$ kg m^{-3} and of conductivity $\lambda = 1.5$ kcal h^{-1} m^{-2} °C^{-1} to obtain a delay of 12 hours then

$$\phi = \frac{24}{2} \sqrt{\frac{\lambda \rho C}{\pi 24} \frac{x}{\lambda}}$$

i.e.,

$$12\,h = \sqrt{\frac{1.5 \times 0.22 \times 2,450}{24 \times 3.14} \frac{x}{1.5}}$$

whence $x = 0.45$ m which corresponds to a conduction speed of

$$0.45 = 12\,h \times v$$

$$v = 3.7 \text{ cm h}^{-1}.$$

The air circulates at rates up to 19 l s^{-1} at the vents, which are 2.2 m apart:

$$\text{flow in l s}^{-1} = \frac{Q_{\text{extracted}}}{\omega_{TS} C_p (t_s - t_e)\, 3,600}$$

where $\omega_{TS} = 1.293 \times \dfrac{273}{273 + t_s}$ kg m^{-3}

Q_{ex} in watts is the usable thermal power (calculated from $Q_{\text{ex}} = G\overline{\tau\alpha} - K(t_m - t_a)$)

C_p = thermal capacity Wh kg^{-1} °C^{-1}

t_s = airflow temperature

t_e = temperature at entry

For example: if $Q_{\text{ex}} = 800$ Wh m^{-2}, $t_s = 50$ °C, $t_e = 17$ °C the flow is

$$\frac{800}{1.09 \times 0.34 \times 3,600 \times 33} = 18 \text{ l s}^{-1}.$$

When the received radiation is at its maximum the temperature of the external wall can reach 70 °C.

The moisture contained in the concrete walls is eliminated by diffusion but as this takes several years performance is affected. It would be better to use either a concrete block construction with a plastic coating to enhance the absorption of incident radiation, or to use one of the concretes developed for road surfaces.

In order to improve the regulation of this system it can be fitted with a control system linked to the supplementary heat supply (see Chapter 6 'Practical Examples').

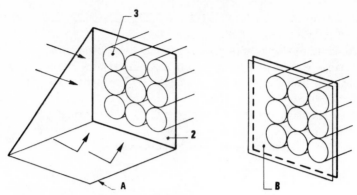

Figure 98. Storage in water canisters.

A. Shutter lowered during day in winter and during night in summer. 1. Aluminium, 2. glazing, 3. cylinder of water.
B. Shutter raised during night in winter and during day in summer

The absence of an external night-time shutter encourages the negative greenhouse effect, which leads to heat losses from the walls to the outside.

Natural ventilation in summer is obtained by air circulating from the north façade towards the collector wall or by a system of openings which allow the hot air to be released.

This system has been applied in solar houses at Odeillo (Basses—Pyrénées) and at Chauvency-le-Château (Meurthe-et-Moselle). The efficiencies are about 50% at Odeillo (1,800 m), and 30—35% at Chauvency-le-Château, where the radiation levels are lower.

(b) *A wall formed of water-filled canisters:* Canisters of water can be used instead of masonry, with a thermal capacity of almost five times that of concrete,
(1.16 Wh kg^{-1} °C^{-1} for water
0.25 Wh kg^{-1} °C^{-1} for concrete.)

An aluminium panel can be used during the winter to increase the radiation by reflection and can also serve as the night-time shutter to reduce the negative greenhouse effect. Aluminium is not very absorbent and its temperature is raised little by the sun.

In summer the panel is lowered at night to cool the water in the canisters, which allows the air circulating through the house by convection to be cooled.

This system has been used by Steve-Baer in New Mexico, U.S.A. (latitude 35 °N).

The great advantage of this approach is that it provides thermal comfort in winter and summer by means of a simple and inexpensive system.

(c) *A collecting roof:* The water is contained in black plastic bags under a transparent film. The principle is similar to that used in Japanese solar water heaters (Figure 89), and a system of covers made from polyurethane insulating panels

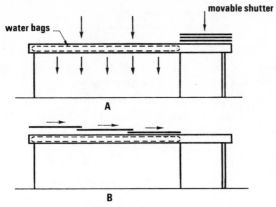

Figure 99. A collecting roof

mounted on rails fulfils the same role as the opening shutter described above.

The panels cover the water containers during the night in the winter and during the day in the summer.

(d) *Collecting wall and adjustable shutter* (ABC Group, Marseilles): Cases b, c, and d are examples of systems using partially adjustable insulation. The use of adjustable shutters as in Figure 100 — provides a wider range of control.

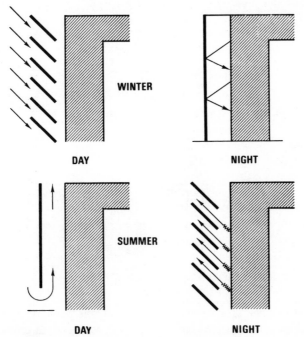

Figure 100. Diagram of a collecting wall using an asjustable shutter

5.2.2 Heat transfer systems, water, or air

Water systems

As we have seen in Chapter 4 and Figure 43, each system contains collectors, storage, distribution, control, and supplementary heating.

The basic circuit is the same as that for a solar water heater but as the requirements are larger the collector surfaces have to be larger. For example, 40 m² is required for a private house of 120 m² floor area as compared to the 6 m² collectors needed for a solar water heater.

(1) *Collectors (see Chapter 4)*   Collectors can be mounted in several ways:

1. in parallel,

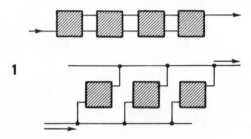

Figure 101. Grouping of collectors in parallel

2. in series,

Figure 102. Grouping of collectors in series

3. in combination,

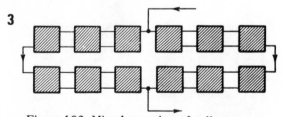

Figure 103. Mixed grouping of collectors

For the same number of units the efficiency and output temperature will differ depending upon whether the units are connected in parallel or in series.

The combined approach allows a better mix of water and a better heat transfer coefficient in each collector.

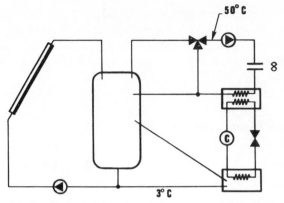

Figure 104. Combination of store and a heat pump

(2) *Storage*   The need to obtain the lowest possible return temperature leads to three possible approaches (see Chapter 4.7 and the preceding paragraph).

(a) two storage tanks:
One hot tank (water having arrived from the collectors), and one cold tank (from which the water is taken into the collectors);
(b) a stratified tank supplying heat directly without an exchanger. This increases the efficiency of the installation (Figures 82 and 95);
(c) a heat pump extracting heat from the return water in the primary circuit (Figure 85) or placed in the distribution network. Distribution of the heat is achieved using a fan-driven water—air convection heat exchanger. The operating temperature is fixed at 50 °C and once the temperature of the store is high enough the circuit operates directly. The return passes through the condenser in the heat pump where it is partially reheated.

According to demand, the fluid can pass back to the store or not, through a three-way valve.

The heat pump evaporator uses as its heat source the cold water from the centre of the tank which it passes to the primary return circuit at a temperature of 3 °C. Here it is mixed with water from the lower section of the store.

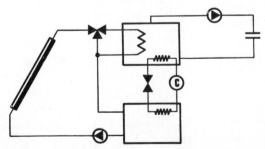

Figure 105. Heat pump connected to two tanks

(3) *A heat pump connected to two tanks*   When the value of incident radiation is high (10 to 14 h in winter) the fluid is directed to the upper tank. Conversely, when the intensity is low (on a day with cloud-covered sky), as the temperature level is lower, water from the collectors is directed to the second tank by a three-way valve.

A heat pump transfers the heat from the lower tank to the upper one.

(4) *Exchange surfaces*   The following can be used:

radiators,
hot air blowers, or
underfloor heating coils.

Radiators: The temperatures reached in the store do not match those produced in a traditional central heating system. Therefore, a permanent supplementary heating system has to be incorporated. The solar radiation is used in this case as a preheater for the water in the distribution circuit.

Convectors: The air in the rooms is heated by low temperature heat exchangers (40 °C). This level of temperature necessitates larger exchange surfaces than for traditional convectors which operate at a minimum of 75 °C.

Underfloor heating coils: This is a conventional system which is used on a large scale in oil-fired or, more recently, geothermal installations (Melun-Sénart, Creil).

The surface temperature of the floor should not be higher than 26 °C (31 °C for bathrooms). The maximum permissible temperature for a concrete tile is 60 °C; the thickness of concrete above the heating network is 0.02 m and the minimum distance from the internal surface of the outside walls is 0.15 m.

The thermal capacity of the floor allows the redistribution of heat after sunset (see Chapter 6 'Practical Examples', Aramon).

(5) *Supplementary heating – Its siting within the system*   In order to meet the whole demand, it would be necessary to have disproportionately large collector surfaces to allow for the non-sunny periods.

It is, therefore, preferable to optimize the collector areas and to complete the installation with a supplementary heating system. It is theoretically possible to

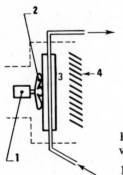

Figure 106. Diagram of the principle of forced convection heater.

1. Motor, 2. fan, 3. heat exchanger battery, 4. shutter

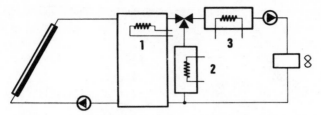

Figure 107. Location of the supplementary heating system

envisage several positions for this equipment, and the choice is made by digital or analogue simulation of the system.

Example No. 1: When placed in the upper part of the store it supplies the energy required to bring the water up to service temperature, but it could heat unnecessary quantities of water if the tank is not well stratified. This system has a long response time and requires several hours to build up the following day's energy supply in the water.

Example No. 2: The primary circuit (the collectors) can be isolated, and the system operates in a conventional manner until the storage temperature reaches the required service temperature.

Example No. 3: The primary system is used for preheating, and when the temperature in the store is too low the three-way valve operates to complete a loop on the auxiliary heating system only. It is then similar to Example No. 2.

(6) *Heating and hot water production* Both applications are frequently grouped in the same installation. The store provides preheated water for an accumulator containing an electrical resistance heater as central heating and domestic hot water require different service temperatures.

Control: See Chapter 4 'Control' and Chapter 6 'Practical Examples'.

(7) *Air systems* In this process the hot air used for heating buildings is not provided by a convector (the preceding system) but is produced either directly

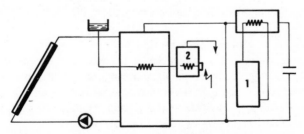

Figure 108. Central heating and hot water production.

1. Supplementary heating system, 2. hot water tank

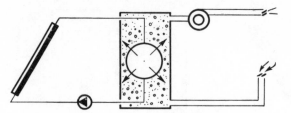

Figure 109. Air system with pebble storage

from the primary circuit or by passing over the pebbles in a combination circuit where water is used to transfer heat from the collector to the store.

Air collectors and pebble storage have been described in Chapter 4 and the system is presented below in a simplified form.

The advantage of this system arises from the fact that the fluid used for heating a dwelling is the same as that already contained within the dwelling. Furthermore, the installation costs for the system are reduced even though the storage volume is greater than required for water. The problems of ice and corrosion do not arise, and there can be no leaks to cause damage to the building.

It is, however, advisable to consider the humidity of the blown air for reasons of thermal comfort.

This system is not recommended in the conversion of existing buildings because of the size of the pebble store. The power consumption of the fans is greater than for water circulation pumps and the preheating of the domestic hot water is much slower than when a water store is used.

In Chapter 6 several examples of applications using this system are shown.

5.3 SOLAR AIR-CONDITIONING

In some climates it is virtually impossible to maintain the internal temperature of buildings at a level compatible with human activity without the aid of artificial devices. The maximum comfort level is about 28 °C.

In the absence of air-conditioning the temperature reached is the result of heat gains caused by the external temperature, insolation, the inertia of the construction and the type of occupancy.

It is desirable to air-condition dwellings in regions where the mean daily temperature goes above the comfort level, for example, in Southern Spain and Sicily, where the average daily temperature during the hot season is 30 °C and over. Only houses with very large inertia can dispense with air-conditioning (wall thickness of 50 cm, flooring of a low thermal conductivity material, such as marble tiles, or a water tank under the houses as at Stromboli).

Therefore, in these regions it is necessary to air-condition houses with average inertia (walls weighing between 100 and 200 kg m^{-2}). However, above 39° latitude air-conditioning is not as essential.

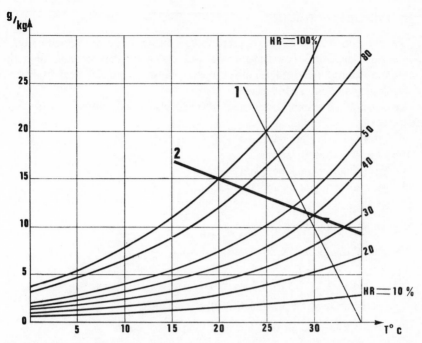

Figure 110. Psychrometric chart (on the abscissa − air temperature, on the ordinate − amount of water vapour per kg of air).

1. Constant effective temperature line, 2. constant enthalpy line

5.3.1 Air humidity

Very little of the water vapour produced by the occupants can be absorbed by air that is already close to saturation point with the result that the body cools little and there is a sensation of excessive heat. Thus, ambient conditions of 26 °C and 80% relative humidity are *felt* to be the same as 35 °C and 25% relative humidity. Both these atmospheres have the same *effective temperature*. The psychrometric chart shows that the comfort level corresponds to an effective temperature of 24 °C with 50% relative humidity.

Humidity problems occur in two types of climate:

a warm humid climate, for example, Abidjan. It is difficult to be comfortable here without air-conditioning because the temperature difference between day and night is only 10 °C; strong fans have to be used to get an airspeed which will increase the convective cooling of the body. The inertia of the construction has little influence in this case;

a hot dry climate. Here the mean daily temperature is above the comfort level. However, the day-night differentials are very large (about 20 °C) and this permits the use of natural air-conditioning (see Section 4.3.3).

5.3.2 Estimation of demand

The amount of heat to be extracted from a building in order to maintain a given temperature depends upon the internal and external temperatures, upon the internal and external relative humidities, upon solar gains and the incidental gains due to occupancy and upon windspeed.

As demand determines the size of the solar installation it is necessary to estimate:

the solar gains through windows, walls, and roof,

the gains due to infiltration,

the gains due to occupancy, and

the gains due to lighting and appliances.

Energy gains through the windows

The gain through the windows can be written

$$q_{windows} = \left(\frac{\alpha K}{h_e} + \tau \right) G + K(t_e - t_i)$$

where α is the absorption coefficient of the pane of glass,

τ is the transmission coefficient of the pane of glass,

G is the received radiation.

Strictly speaking τ is different for each component; G is calculated according to the method in Chapter 3.

Gains through opaque walls

$$q_1 = \frac{K\alpha G}{h_e} + K(t_e - t_i).$$

It is advisable to include the heat transfers caused by heating and cooling of the materials, and to do this a damping factor is introduced, $m < 1$,

where $m = 0$ for heavy masonry,

$m = \frac{1}{2}$ for cavity block wall,

$m = \frac{3}{4}$ for light wall.

If the wall were infinitely heavy, the damping would be complete; the heat flux would be constant and equal to

$$q_2 = K \left(t_{em} + \frac{\alpha G_m}{h_e} - t_i \right)$$

where G_m is the average value of G,

and t_{em} is the average value of t_e.

For a real wall the flux fluctuates around q_2:

$$q = K\left(t_{em} + \frac{\alpha G_m}{h_e} - t_i\right) + mK\left(t_e + \frac{\alpha G}{h_e} - t_{em} - \frac{\alpha G_m}{h_e}\right)$$

$$q = \underbrace{K(t_{em} - t_i) + mK(t_e - t_{em})}_{\text{transfer due to temperature variation}} + \underbrace{\frac{K\alpha G_m}{h_e} + \frac{\alpha m K}{h_e}(G - G_m)}_{\text{transfer due to solar radiation}}$$

transfer due to temperature + transfer due to solar
variation   radiation

Example: South of France,
$t_{em} = 26°$, $t_i = 26°$, external temperature $t_e = 32°$.
Heat flux through a heavy vertical wall $m = 0.2$ of average insulation $K = 1.74 \text{ W m}^{-2} \text{ °C}^{-1}$:

$\alpha = 0.2$,

$h_e = 18$.

(1) caused by temperature variations:

$K(t_{em} - t_i) = 0$, $mK(t_e - t_{em}) = 2.32 \text{ W m}^{-2}$,

(2) caused by solar gains:

$\dfrac{\alpha \, mK}{h_e} = 7.73$, $\qquad\qquad G = 750 \text{ W m}^{-2}$ at 12.00 h,

$\dfrac{\alpha \, KG_m}{h_e}(G - G_m) = \dfrac{1.35 \text{ W m}^{-2}}{9.08 \text{ W m}^{-2}}$, $G_m = 400 \text{ W m}^{-2}$,

giving a total of 11.14 W m^{-2} between 12.00 and 13.00 h. Gains caused by air infiltration are:

$0.35 \, NV(t_e - t_i)$.

Internal gains

A seated person generates between 80 and 90 W per hour and adds about 50 g per hour of water vapour to the ambient air.

These gains should be determined for each hour in order to find the maximum amount of heat to be extracted (usually between 3 and 4 o'clock in the afternoon).

5.3.3 Systems used in solar air-conditioning

Solar energy is economical in a climate requiring both air-conditioning and central heating as the installation is in service without interruption, and in this case the air-conditioning function will be needed during the period of peak radiation.

There are three main categories of system: those using a heat pump, those using the absorption principle and those making use of natural phenomena.

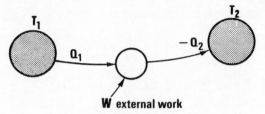

Figure 111. The principle of the heat pump

Heat pump systems

The chief benefit of this method is that the same piece of equipment is able t supply heating and cooling cycles.

Heating cycle   The heat pump transfers heat from a *cold* source to a hot sin through the addition of external work.
Its coefficient of performance is

$$COP = \frac{-Q_1}{W} \text{ always} > 1.$$

A COP $\geqslant$ 2.5 is aimed at in order to regain the initial energy supplied to th power station, which has an efficiency of about 40%. A *cooling cycle* transfers he; from the cold body and

$$\eta = \frac{Q_2}{W} = \frac{T_2}{T_1 - T_2}$$

The physical principle underlying the cycle is that liquids under high pressu boil at a higher temperature than liquids under low pressure. For example, Freo 12 is used as a heat—transfer fluid.
There are two possible methods of operation:
the solar heating system and the heat pump can be independent of each other, c
the solar heating system and the heat pump can be coupled, providing a mo effective solution.
The solar heating system can be used to increase the coefficient of performanc of the heat pump. In effect, the energy stored in the form of sensible heat in th storage tank reduces the work of the compressor which has to raise the temperatu of the refrigerant to the desired temperature (see Figures 104 and 105).
A heat pump is particularly appropriate for use when the differences betwee the mean summer and winter temperatures are not large, as the heating and coolin capabilities are essentially the same.
Air to air heat pumps must be capable of supplying the heat demand at a outside temperature of −4 °C without using any supplementary heating. As th outside temperature decreases the coefficient of performance falls.
An additional resistance heater (4 kW) is sufficient to ensure the heat requir

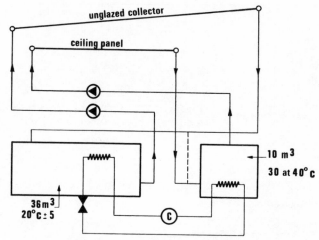

Figure 112. Diagram of installation

ments and complements the heat pump output until the outside temperature falls to −11 °C (for a well insulated house of 150 m²).

Example of use of a heat pump in a Japanese house latitude 36 °N. (See also the Philips house)

The water collectors are on the roof, inclined at 15°. They have no coverplate and storage is in two underground tanks. See Figure 112.

Heating. The water from the collectors circulates in a primary tank at 20 °C ± 5 °C. This provides the cold source for the heat pump, which raises the temperature of a second small tank to between 30 and 40 °C. The radiant ceiling panels which heat the rooms are supplied from this small tank.

The power of the heat pump compressor is 2.2 kW.

Air-conditioning. The water in the small tank which was warmed up during the day cools down at night by radiation and convection as it circulates through the rooftop collectors. During the day the heat pump takes up heat from the water in the large tank and transfers it to the small tank. The distribution circuit is linked to the large tank whose temperature is at 10 °C ± 5 °C.

There is no cycle reversal − the heat pump always operates in the same direction and acts as a transfer mechanism. At night the roof is used as an evaporator in this installation which is based on natural air-conditioning.

Absorption Air-conditioning systems

Solar air-conditioning can be effected either by compression machines or by absorption machines.

Refrigeration necessitates the removal of a quantity of heat from the space being cooled. According to the second law of thermodynamics, external work must be supplied to effect this transfer. The warm body receives the heat extracted from the cold body and the heat equivalent of the work expended.

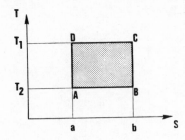

Figure 113. Cycle of a refrigerator on a temperature–entropy chart (T,S)

The cycle of a refrigerator is therefore that of a reversed heat engine. Th performance of a refrigeration installation is measured by the coefficient of pe formance for refrigeration:

$$COP_r = \frac{Q_2}{W} = \frac{Q_2}{Q_1 - Q_2} = \frac{T_2}{T_1 - T_2}.$$

Q_2 is the quantity of heat extracted from the cold source at temperature T_2 = are $aABb$,
Q_1 is the quantity of heat supplied to the hot sink at temperature T_1 = area aCD

The cycle described in Figure 114 is the inverse of the Rankine cycle whi applies to steam engines and where
AB is evaporation,
BC or BC' is an adiabatic expansion with partial condensation,
CD or C''D is the condensation to saturated liquid at D.

$$\text{Efficiency} = \frac{\text{area ABCD}}{\text{area ABcd}},$$

since 1 KWh = 860 kcal we can use the refrigeration effect

$$K_f = 860 \times COP \text{ kcal kWh}^{-1}.$$

The absorption refrigerator Instead of using a motor compressor one can us quantity of heat Q_3 from an auxiliary source at a temperature T_3 higher than T
The only difference from the vapour compression machine is the method us for compressing the refrigerant liquid, and the temperature variation of the so

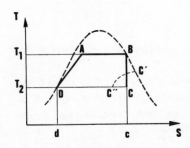

Figure 114. Rankine cycle on a temperature entropy chart (T,S)

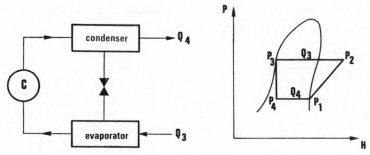

Figure 115. Vapour compression refrigeration cycle

bility of a gas in a liquid is used. Water can absorb up to 900 times its own volume of gaseous ammonia NH_3.

It should be noted that the evaporator E, the condenser L and the valve R are still present. The compressor is replaced by a mechanism which operates as follows:

In the liquid state NH_3 enters the evaporator E where it removes Q_2 from the refrigerated space.

The NH_3 vapour which results from the absorption of Q_2 passes to the absorber A where it releases a certain amount of heat by dissolving in water at a low temperature.

The concentrated solution then passes into the generator G (distillation vessel) where it is heated: the NH_3 is boiled off and passes to the condenser L where it liquifies, releasing Q_1, and continues its cycle by passing through the throttling valve to the evaporator E where the pressure is lower, allowing it to be vaporized by absorbing the amount of heat Q_2 from the refrigerated space.

The dilute solution left in the generator returns to the absorber A.

Expression for the coefficient of performance L and A are assumed to be at

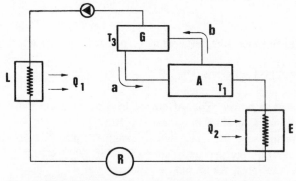

Figure 116. Diagram of an absorption refrigerator.

E. Evaporator,   A. absorber,   G. generator,   L. condenser, R. flow valve.
(a) Diluted solution, (b) concentrated solution

temperature T_1. They receive an amount of heat Q_1 from the evaporator (Q_2) a from the generator at temperature T_3 (Q_3). Thus,

$$Q_2 + Q_3 = Q_1.$$

If the cycle is reversible,

$$\frac{Q_2}{T_2} + \frac{Q_3}{T_3} = \frac{Q_1}{T_1} = \frac{Q_2 + Q_3}{T_1}$$

and the coefficient of performance

$$\text{COP} = \frac{Q_2}{Q_3} = \frac{T_2}{T_3} \times \frac{T_3 - T_1}{T_1 - T_2}$$

$$= \frac{\text{the refrigeration effect in the evaporator}}{\text{the heat absorbed by the solution during regeneration}}$$

For example, if $\theta_3 = 100\,°\text{C}$, $T_3 = 273 + 100 = 373$ K

$$\theta_1 = 20\,°\text{C}, T_1 = 293\,°\text{C}$$

$$\theta_2 = -10\,°\text{C}, T_2 = 263\text{ K}$$

$$\text{COP} = 2.$$

In practice, however, conditions are not as favourable as this because of irreversibilities in the cycle and COP < 1.

Numerical example. The aim is a refrigeration effect of 100,000 kcal h^{-1}

condenser temperature = 30 °C,

evaporator temperature = −5 °C,

generator temperature = 100 °C,

absorber temperature = 38 °C,

concentration of the dilute solution leaving the generator $m_1 = 0.340$,

concentration of the concentrated solution leaving the absorber $m_3 = 0.440$.

The flow of ammonia. The tables for ammonia vapour show that each kilogram of refrigerant transferred between +30 °C and −5 °C produces:

$$400.2 - 133.8 = 266.4 \text{ kcal.}$$

It is necessary to produce 100,000 kcal h^{-1}, therefore, the flow will be

$$\frac{100,000}{266.4} = 375.3 \text{ kg h}^{-1} \text{ of NH}_3.$$

Theoretical pumping power. The difference in concentration between the conc trated and the dilute solution is $m_2 - m_1 = 0.10$.

10 kg of solution must be circulated to ensure 1 kg of NH$_3$ in the refrigerat circuit. The circulation pump must ensure a flow of 375.3 × 10 = 3,753 kg h^{-1}.

According to the tables for ammonia vapour

$t = 30\,°\text{C}$ (at the condenser), pressure = 11.895 atm, which is also the pressur the generator, and at

$t = -5\,°\text{C}$ (evaporator), the pressure = 3.62 atm.

The circulation pump must, therefore, deliver an hourly flow of 3,753 through a height of $(11.895 - 3.62) \times 10.33$ m (1 atm = 10.33 m of water) $\simeq 85.48$ m, therefore, *the theoretical output of the pump* is

$$\frac{3{,}753 \text{ kg} \times 85.5 \text{ m}}{3{,}600 \text{ s} \times 75} = 1.18 \text{ hp} = 0.86 \text{ kWh} \ (1 \text{ hp} = 75 \text{ kgm s}^{-1}).$$

Let us compare this with an ammonia compressor of the same output. The work of compression would be 40 kcal kg^{-1} and the theoretical power

$$P_{th} = \frac{40 \times 375.3}{860} = 17.4 \text{ kW} \ (23.7 \text{ hp}).$$

Use of solar energy in an absorption refrigerator

Two types of absorption—process refrigerator are currently manufactured: one operates with a mixture of ammonia and water (NH_3-H_2O), and the second operates with a mixture of lithium bromide and water ($LiBr-H_2O$).

In the first type of machine, water is the absorber and NH_3 is the refrigerant. In the second type of machine, water is the refrigerant and the lithium bromide is the absorber.

In an NH_3-H_2O system there must be a rectifier column which prevents the water vapour mixed with NH_3 from entering the evaporator where it could freeze. This does not occur when $LiBr-H_2O$ is used because water is the refrigerant. Further the NH_3-H_2O system requires higher operating temperatures in the generator, between 120 and 150 °C whereas $LiBr-H_2O$ operates in the range 88 to 93 °C.

In order to reach these temperatures it is necessary to use *either* collector plates

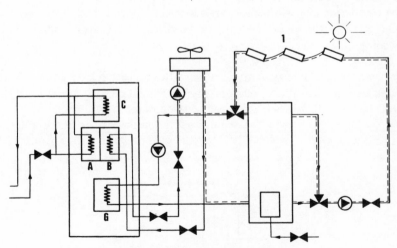

Figure 117. Example of solar air-conditioning.

––––– heating; ——— air conditioning.
1. Collector plates;
G. generator, A. absorber, C. condenser, B. evaporator

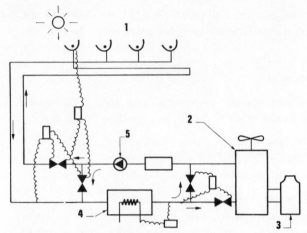

Figure 118. Example of solar air-conditioning with cylindrical parabolic collectors.

1. Collectors, 2. refrigeration, 3. cooling tower, 4. auxiliary heating, 5. pump

with a selective absorber and a double glazed coverplate (as the efficiency drops for high temperatures the collector surface will be bigger for air-conditioning than for heating (see Figure 61), or parabolic cylinders with a low concentration factor (5), with some tracking facility, as in Figure 75.

Temperatures of 100 to 130 °C above the ambient air temperature can be reached with efficiencies of 40 to 50%.

The coefficient of performance varies between 0.5 and 0.85. The collectors are inclined at the latitude of the site (or at latitude $-12°$ whichever is the more favourable).

Air-conditioning by evaporation   This method is particularly well suited to hot dry climates. The principle consists of passing the external air over a moist surface or through a spray before being supplied to the rooms.

The cooling produced by the water evaporation lowers the temperature of the air as well as raising the humidity. On the psychrometric chart (Figure 110) the

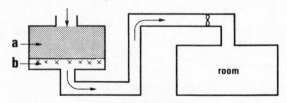

Figure 119. Schematic diagram.

a. Rocks, b. water spray

point is displaced along a constant enthalpy line corresponding to a change of state with no energy gains or losses.

A variation on this principle consists of evaporating water on the insides of the walls. The walls are thus cooled and, indirectly, so is the air of the building without actually increasing its humidity. This is the control principle used in termitaries (see Chapter 3).

Another variation uses the passage of air over crushed rocks.

Cooling of intake air Fresh air from outside the building is drawn through a water spray in the storage vessel. The air is cooled to near the saturation temperature, and the entire storage contents are lowered to this temperature after several hours. The next day when air-conditioning is required the external hot air is drawn through the store where it is cooled to room temperature by evaporation. This system is used in Australia and the U.S.A., and the Thomason house described in Chapter 6 illustrates one example.

5.4 SOLAR DISTILLATION

This process offers one solution to the problem of supplying fresh water, and is applicable for cases where the demand is limited (200 m^3 per day) and where the climate is suitable.

The most common still is based on the greenhouse effect and there are many variations in use.

5.4.1 Principle of operation

The solar still reproduces the earth's water cycle on a small scale. The solar radiation reaching the surface of rivers, lakes and oceans is largely absorbed as heat, evaporating water in the process. The vapour thus produced is mixed with air which is moving under the action of the winds.

Later, when the combination of air and vapour is cooled to the dew point, condensation can occur and fresh water precipitates in the form of rain.

A still comprises a limited area of salt water with a transparent covering which

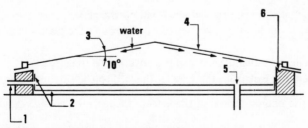

Figure 120. Diagram of solar still.

1. Salt water intake, 2. casing of the tank, 3. pitch of covering (10–12°), 4. glazed covering, 5. brine outlet, 6. outlet gutter for condensed water

prevents the humid air from escaping and which also forms a cold surface on which the vapour can partially condense. This covering forms a screen to infrared radiation emitted by the surface of the water.

As the cover is at an angle of inclination not less than $10°$ in order to receive the maximum radiation, it also prevents the distilled water which flows by gravity down the inside of the covering, from being mixed again with the salty water.

The tank which contains the salt water is watertight and painted black in order to absorb the solar radiation well. The distilled water is collected in gutters at the base of the slope of the covering

In solar distillation the daily output of distilled water is variable as it is dependent upon the intensity of solar radiation. Therefore, it is at its maximum during the day but will be reduced under cloudy skies and it becomes zero at night. These short and long term variations necessitate a store for the distilled water as water demand is generally uniform.

5.4.2 The output of solar stills

On a good day: 4 1 m^{-2},
in winter: 1.25 1 m^{-2} for clear periods, and
in summer: 4.90 1 m^{-2} for clear periods.

Annual production is of the order of 1 m^3 m^{-2}. An average demand of 19 m^3 per day corresponds to a still with an area of 7,000 m^2. There are numerous examples of solar stills in Australia, Tunisia, and Greece, where the largest solar still is in service on the island of Patmos. This still is 8,600 m^2, has a glass cover and was built in 1967.

5.4.3 Energy balance

The direction and intensity of radiation on the surface of the still are constantly changing; the ambient temperature and the wind also change constantly. Therefore, this is a dynamic system and one can write

$$\alpha_{glazing}G + \alpha_{water}\tau_g G = q_1 + q_2 + C_p \frac{\partial t_e}{\partial \theta}$$

where $\alpha_{glazing}G$ is the solar energy absorbed by the glass,

$\alpha_{water}\tau_g G$ the solar energy transmitted through the glass and absorbed in the tank,

q_1 the energy transmitted from the covering to the air,

q_2 the energy lost from the base of the still to the surroundings,

$C_p \frac{\partial t_e}{\partial \theta}$ the energy stored in the system through the change in water temperature with time θ.

One can also write

$$q_1 = q_r + q_c + q_e + \alpha_g G,$$

123

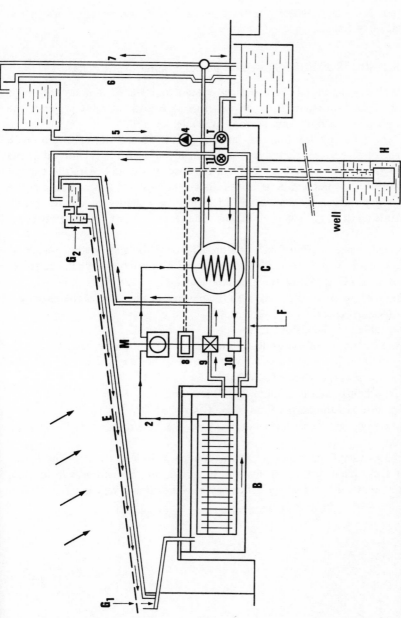

Figure 121. Diagram of the principle of a solar engine.

1. Oil circuit, 2. gas circuit, 3. pumped water circuit, 4. control valve for water flow, 5. output pipe, 6. overflow, 7. supply from tank, 8. water pump compressor, 9. oil circulation pump, 10. reinjection pump, 11. auxiliary circulation pump.

G_1 Collecting gutter, G_2 distributing gutter, E oil flow, T turbine

where q_2 is the exchange through radiation from tank to cover, q_c the exchange by convection from tank to cover, q_e the exchange of energy by evaporation–condensation.

The output is $R = q_e/L$ where L is the latent heat of the vaporization of water at the temperature of the still.

5.5 SOLAR ENGINES (used for pumping water in arid zones)

To reach their full potential these devices should be installed in areas with high, regular year-round insolation. After an experimental phase in pilot projects, the system described here is now commercially available.

One of the first applications was at Chinguetti, Mauritania. The flat plate collectors are of the trickle-flow type, using locally available vegetable oil as the heat transfer fluid. The collectors also form the roof of a school. With this type of collector, however, output temperatures are not very high and this necessitates the addition of a high efficiency gas motor which must be reliable and able to operate without maintenance despite sand problems. The useful area of collector surface is 70 m².

The pumped water at Chinguetti is stored in two fibreglass cisterns, each of 15 m³. The apparatus and the tanks are housed in a tower from which the water is distributed to drinking fountains and animal troughs.

The design had to be in harmony with the mosque at the oasis, as the mosque is one of the oldest in Africa.

The water table is at a depth of 15 m.

The maximum output of the wells is 10 m³ h⁻¹ for 5½ hours per day.

The consumption per inhabitant is low (10 to 15 l per day).

The pump supplies the needs of 3,000 people.

The gas motor turns at a maximum of 220 revs/min.

The temperature of the engine fluid is 40 °C $< t <$ 55 °C.

The water temperature varies from 73 °C in the boiler to 30 °C in the condenser.

The pressures vary from 4 kg m⁻² in the condenser to 7 kg cm⁻² in the boiler.

Some of these plants are to be installed in agricultural cooperatives in Mexico, but it must be remembered that the efficiency of this engine is very low (only a few per cent).

6

Practical Examples of Solar Buildings

One of the first solar houses was built in 1939 at M.I.T., Massachusetts, and since then many others have followed. We will describe some of these houses and give schematic diagrams.

Each example exhibits an ingenuity in design of collectors and storage, and the control systems are particularly effective while based on simple components — thermocouples, Wheatstone bridges, differential temperature detectors, relays to switch the circulation pumps in the collector circuit and to trigger the auxiliary heating circuit.

The nature of the climate determines the solar application adopted, for instance, in our moderate latitudes air-conditioning is not normally required except in special circumstances.

One French house in the course of construction has been studied by G.R.E.E.N. (Groupe de Recherches et d'Etudes sur les Energies Nouvelles, 32 rue de Lille, 75007 Paris; 108 rue Chevreul, 92 Nanterre) and supplies the necessary information on the procedure to be followed.

6.1 M.I.T. SOLAR HOUSE NO. 4, MASS., U.S.A.

Built in 1954–55. Designed by Department of Mechanical Engineering, M.I.T.

Situation

Latitude 42° North
Longitude 71° West
Degree-days: 3,300
Temperature minimum −13 °C
Average energy received in January: 1,860 Wh m^{-2} per day

Building

Two storeys
Area: 134 m^2
Energy demand: 19,617 kWh (October to March)

125

Figure 122. M.I.T. solar house No. 4

Aim

To meet 75% of the demand by solar heating

Collectors

Integrated into the building shell
Type: water
Position: south facing roof
Inclination: 60°

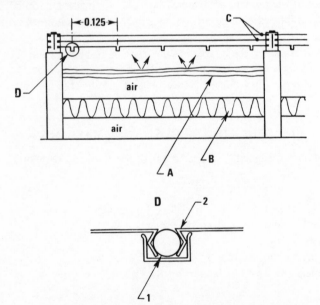

Figure 123. Cross-section of the solar collector.

A Aluminium sheets, B 7.5 cm of insulation,
C glazings, D detail ((1) copper pipe, (2)
aluminium painted black)

Area: 60 m²
Construction: aluminium sheets clipped to copper tubes
Glazing: double glazed

Storage

Type: water
Volume: 6 m³
Thermal capacity: 6.5 Wh °C⁻¹
Temperature: 49 °C

Heating

Water flow in the collector circuit: $2,430\,1\,h^{-1}$ $(40\,1\,h^{-1}$ per square metre of collector)
Water flow in the distribution circuit $817\,1\,h^{-1}$
Heating system: blown air heaters, $0.4\,m^3\,s^{-1}$
Motors: 2 circulation pumps + 1 fan
Supplementary heating: oil
Power: 12 kW
Heat transfer system: by a heat exchanger in the thermal store

Characteristic: The store separates the primary circuit from the heat distribution circuit.

Domestic hot water: The water is preheated by passing through the primary storage tank and it is then raised to the service temperature in the second tank.

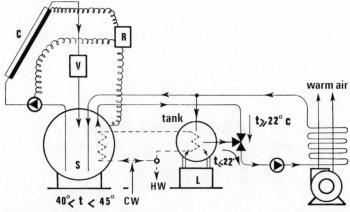

Figure 124. Diagram of installation.

C Collectors, R primary circuit control, V expansion vessel,
S store, CW cold water, HW hot water, L supplementary heater

Performance

    Average efficiency: 40.8 to 45%
    Solar contribution: 11,163 kWh (13,880 including hot water) which represents a
    saving of between 48.2 to 56.6%

Control by normal resistance thermometers in the collector and the store. The thermometers are connected in two arms of an alternating current Wheatstone bridge. Switching is controlled by an electrical relay which is activated by the out-of-balance voltage in the Wheatstone bridge.

6.2 THE DENVER SOLAR HOUSE, U.S.A.

Built in 1956–57 by Saint-Gobain, U.S.A., as part of a solar energy development programme. Designed by Löf.
    This house uses air collectors in conjunction with a pebble store and a natural gas supplementary heating system.

Situation

    Latitude: 40° North
    Altitude: 1,800 m.
    Degree-days: 3,338
    Temperature minimum: −18 °C
    Average energy received in January: 2,550 Wh m^{-2} per day
    The windows are used extensively in this application; those on the south façade are protected by a roof overhang to reduce the summer gains but not those in winter.

Building

    Number of heated storeys: 1½ (195 m^2 groundfloor + ½ the basement 102 m^2)
    Floor area: 297 m^2
    Winter demand: 56,900 kWh

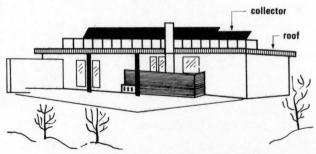

Figure 125. The Denver solar house, U.S.A.

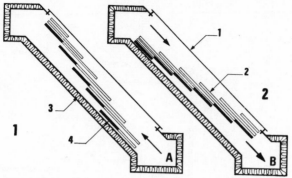

Figure 126. Collectors with cold air (1) and hot air (2).

A Entry of cold air from the shared duct, B. hot air exit through the shared collection duct.
1. Coverplate,   2. glass sheets,   3. insulation,
4. Blackened glass

Collectors

Type: air (see Figure 126)
Position: free standing on the roof
Inclination: 45°
Area: 56 m²
Construction: the absorber is formed from overlapping glass sheets
Glazing: ½ single glazed, ½ double glazed
The collector surfaces are arranged in two banks surrounded by an acroterion. Each bank contains ten cold air panels and ten hot air panels arranged alternately. The cold air from the house or from the two storage units enters by the cold air intake duct and is then distributed through the ten cold air panels in each bank.
After this preheating in the cold panels the air flow towards the hot panels through an intervening system of air ducts.
The air is further heated in the hot panels: one collection pipe and a network of ducts distributes the hot air to the house and the store.
When the collectors produce no heat the store is connected to the distribution system through an outlet at the bottom of the storage cylinders.

Storage

Type: granite stone crushed to 40 mm average diameter
Volume: 7 m³
Location: centre of the house
Tank: 2 cylinders of diameter 0.90 m and height 5.40 m
Temperature: 60 °C
A 0.27 m pipe through which the air flows while giving off its heat runs from the top to an opening at 0.60 m from the bottom.

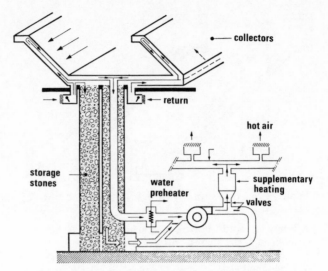

Figure 127. Cross-section of the heating system

Heating

The air is blown through the ducts by a two speed 1½ hp fan.
The airflows are

(a) in the primary circuit: 1,300 m³ h⁻¹
(b) in the distribution circuit: 1,300 m³ h⁻¹ (to the occupied rooms)

supplementary heating: natural gas
power: 30 kW
linked to the system by heating the air from either the collectors or from the
store (see diagram)
gas consumption: 7,181 m³

Domestic hot water: This is preheated in an air-water heat exchanger in front of
the fan on the main duct. A small portion of the air heat is transferred to the water
circulating by thermo-siphon from a 300 litre tank.

Control

This controls three cycles:

heat output from the collectors,
storage of heat,
heat output from the store.

The control principle is identical to that of the M.I.T. house with an additional
fine control in each room, using a room-temperature thermostat and adjustable

laps to progressively open or close the ducts according to the room temperature ndicated by the thermostat.

Thus, as soon as the temperature falls below that set by the thermostat this acts imultaneously on the adjustable vent in the room, on the main flaps in the primary ircuit and on the circulating fan.

The thermostat has a second contact which signals to the auxiliary heating ystem if the temperature continues to fall (after 10 minutes).

Performance

Average efficiency of the collectors: 34.6%

Solar contribution: 15,160 kWh = 26.5%

Ratio $\dfrac{\text{area of air collectors}}{\text{area to be heated}} = \dfrac{56}{293} = 0.19$

Electricity expenditure for operation of the system (fan, motors for shutters) nd control instrumentation: 3,876 kWh.

6.3 THE SWEDISH ROYAL ACADEMY BUILDING IN CAPRI, ITALY

Built in 1960 for the Swedish Astrophysical Station in Capri.

Situation

Latitude 41° North
Altitude: 240 m
Degree-days: 1,500
Design temperature: −1 °C

Building

2 storey laboratory
Useful area: 180 m²
Winter demand: 15,256 kWh

Figure 128. Solar building in Capri, Italy

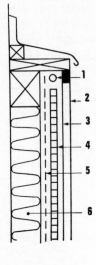

Figure 129. Cross-section of collector.

1. Collector, 2. external glazing, 3. plastic sheet (Testar),
4. 92 x 108 cm radiator, 5. aluminium foil, 6. insulation

Collectors

Type: water
Orientation: south-west due to the effects of shade caused by the relief (see
   Figure 32)
Inclination: 90° (vertical)
Area: 30 m²
Construction: commercial Swedish central heating radiators
Coverplate: 1 external glass and 1 plastic (Tedlar)

The radiators are painted black. A layer of aluminium foil is placed between the
radiators and insulation to prevent water vapour diffusion caused by condensation
on the inner pane.
   This foil also reflects the infrared radiation originating from the radiators.

Storage

Type: water
Volume: 3 m³
Tank: steel
Location: underground
Temperature: 35 to 58 °C

Heating

Total water flow in the primary circuit: 480 l h⁻¹ (16 l h⁻¹ per square metre of
   collector
Total water flow in the distribution circuit: 480 l h⁻¹
Room heaters: radiator panels

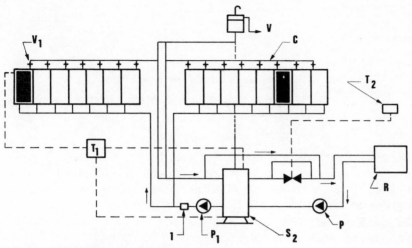

Figure 130. Diagram of installation.

S Storage, P_1 pump $8\,1\,min^{-1}$, P_2 pump $8\,1\,min^{-1}$, T_1 primary control, T_2 room thermostats, R radiator, C collector, V expansion vessel, V_1 control valve, I non-return valve

Motors: 2 pumps
Supplementary heating: stove + separate electric radiator

Control

An array of thermostats (collectors, store, rooms), three-way valves and two pumps control the water distribution.

Domestic hot water

Supplied by a separate collector independent of the system.

Performance

Solar contribution: 70%

6.4 THE THOMASON HOUSES, U.S.A.

These three dwellings were erected between 1959 and 1963. The system supplies air-conditioning and central heating. The main concern of the designer was to produce an economical approach by studying the different parts of the system simultaneously.

An improved trickle-flow collector was developed which was 50% cheaper than the normal type. Its principle of operation is shown in Figure 131.

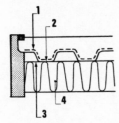

Figure 131. Cross-section of collector.

1. Plastic, 2. water layer, 3. blackened aluminium,
4. insulation

The advantage of these houses is that both heating in winter and cooling in summer are provided by a very simple system in which a single pump circulates the water from the store to the top of the roof collectors and a fan circulates air throughout the house.

The complex and costly components usually found in air-conditioning and heating systems based on blown air are not used in this system.

The extra initial cost of this installation arises from the pebble store (50 t + 6.4 m^3 of water), sited in the basement of the house.

Situation

Washington, D.C., U.S.A.
Latitude: 40 °N
Altitude: 15 m
Degree-days: 2,390 (reference temperature t = 18 °C)
Average energy received in January: 1,850 Wh m^{-2} per day
Minimum temperature: −10 °C
HOUSE NO. 1
Area: 100 m^2 on a single storey

Collectors

Type: trickle-flow
Orientation: south, wall and roof
Inclination: 60° and 45° respectively
Area: 78 m^2
Construction: aluminium sheet
Coverplate: 1 plastic (Mylar) and 1 external glass
Insulation: at the back

Storage

Type: water and pebble
Volume: 6.4 m^3 water and 50 tonnes of pebbles
Tank: one steel tank for the water and a concrete casing for the pebbles
Location: in the basement

Total thermal capacity: 470 kWh
Temperature: 52 to 57 °C

Heating

(a) Primary collector circuit. The trickles of water are heated as they flow down through the collector. The water is then collected in a gutter at the base of the wall and drawn to a primary storage tank which contains a heat exchanger for producing domestic hot water. The water then passes into a 6,400 litre cylindrical tank where it transfers part of its heat by conduction to the mass of stone surrounding it. The cooled water coming from the base of the tank is returned by a small circulation pump to the top of the collector to be reheated.

Flow in the primary circuit: $1,623 \, 1 \, h^{-1}$, or $20 \, 1 \, h^{-1}$ per square metre of collector.

(b) Secondary circuit, distribution. Air from the occupied rooms is heated by contact with the pebbles and is circulated by a thermostatic fan. The air then heats the house through a distribution network sited at floor-level. If the stored heat is insufficient a thermostat brings an auxiliary oil-fired heating system into operation.

Air-conditioning

The basic principle depends upon cooling of the water at night by radiation to the sky, air convection in the pebbles and the evaporation of water.

Instead of being directed to the collectors at night the water in the primary

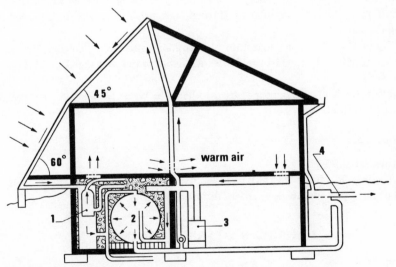

Figure 132. Principle of operation — heating.

1. Domestic hot water, 2. water, 3. supplementary heating, 4. rainwater overflow

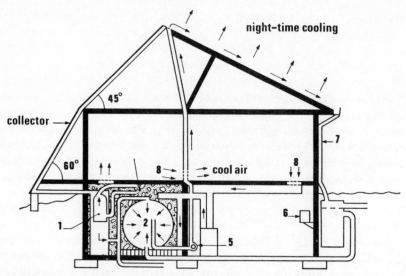

Figure 133. Principle of operation – air-conditioning.
1. Domestic hot water, 2. water, 5. pump, 6. humidifier, 7. return flow from roof, 8. reheated air

circuit is conducted to the north side of the roof where it is cooled by radiation to the sky, by evaporation and by conduction to the cold night air.

The water is then recovered in a gutter and led to a recovery sump where rainwater is added to top up for evaporation losses.

The water returns by gravity to the tank, which in its turn cools the stones surrounding it. The air circulating through the pebbles is used to cool the house during the day.

A humidifier helps to maintain the ambient humidity. During the day the collector circuit only runs to supply domestic hot water and the flow to the store is shut off.

The surplus heat in summer is used to heat the water for a small swimming-pool containing 7,500 litres.

Performance

Estimated at 95%

The cost, at about £1,000, was equivalent to the cost of a conventional installation to provide heating, air-conditioning and hot water.

6.5 THE PHILIPS EXPERIMENTAL HOUSE, 1975

Description

This is an example of a four-person dwelling designed after a very detailed analysis of the different parameters.

This project attempted to limit maximum demand by using very high insulation for the walls and windows and also a heat reclaim system on the extract air and waste water from the bathroom and kitchen.

The demand, thus reduced to a minimum, is met by a circulating water system, made up of an area of collectors on the south-facing roof and a large capacity storage tank for year-round requirements.

Control of the system is achieved by using two mini-computers. The integrated system also uses the ground heat (7 °C) which is transferred by a heat pump to a domestic hot water tank (50 °C) separate from the main store. Summer air-conditioning is supplied by air circulation from the north wall along the walls of the basement through a porous structure, using the ground as storage.

The data logging system records the losses due to each individual element of the structure. The efficiencies of the heat-recovery units on the extract air and the waste water can thus be determined. The efficiency of the collector as a function of service temperature and climatic conditions is calculated and stored on magnetic tapes together with inside and outside temperatures and temperature profiles in the storage units and the soil.

Situation

Latitude: 50° North
Aix-la-Chapelle, France.

Building

Floor area: 116 m²
Volume: 290 m³
Window area: 23.5 m²
Window frames: $K = 2$ W m^{-2} °C^{-1}
Double glazing: $K = 1.9$ W m^{-2} °C^{-1}

The low value of conductivity for the windows is obtained by the introduction of a rare gas (krypton) between the two glass panes and by coating the internal pane with a reflective layer of indium oxide. NOTE: with Eliotherm produced by Saint-Gobain and an air layer of 12 mm, $K = 1.7$ to 1.9 W m^{-2} °C^{-1}.

Figure 134. Philips house

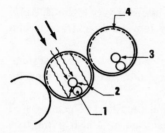

Figure 135. Collector.

1. Hot water return, 2. silvered mirror, 3. black glass tube, 4. deposit of indium oxide

Heating demand: 8,300 kWh per year
Recovered heat: 6,895 kWh per year

Efficiency: $\dfrac{6,895}{8,300} = 83\%$

Collector

Type: water. See Figure 135

The outer protective shield is a commercial fluorescent lighting tube coated with layers of indium oxide to reflect the infrared radiation emitted by the water pipe.

Orientation: south
Inclination: 48°
Area: 20 m^2

Storage

Long term:
Type: water
Volume: 42 m^3
Location: in the basement
Insulation: 25 cm of rock wool
Temperature: 5° to 95 °C
Domestic hot water tank:
Volume: 4 m^3
Insulation: 25 cm of rock wool
Temperature: 45° to 55 °C
Waste water tank:
Volume: 1 m^3
Insulation: 10 cm of rock wool

Storage by the ground

A heat exchanger occupies an area of 150 m^2 underneath the foundations of the house. It is formed by a plastic tube, 120 m long, through which water circulates.

This heat exchanger forms one of the cold sources for the heat pump, which extracts heat and transfers it to the hot tank for domestic hot water. The second cold source is the waste water tank which is also fitted with a heat exchanger.

Heat pump power: 1.2 kW
COP: between 3.5 and 4 over the temperature range $7°-50°C$

Heating

By radiators fed by water from the long-term store.
Air change rate: $300-600 \text{ m}^3 \text{ h}^{-1}$ (1 to 2 air changes per hour) 90% of the heat is recovered from the air extracted by controlled mechanical ventilation.

Air-conditioning

Air is taken from the north face of the roof and at ground level: natural air circulation is used.

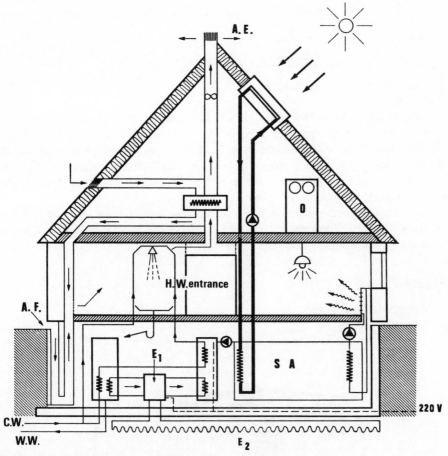

Figure 136. Diagram of operation.

O Mini-computer, SA year round store, E_1 heat exchanger, E_2 soil heat exchanger, AF fresh air, CW cold water, WW waste water, AE extracted air, HW hot water

Figure 137. Aramon house, Group I

6.6 HOUSES AT ARAMON, FRANCE, 1975

Erected for Electricité de France
Architect: G. Chouleur

The project includes two groups of private houses:

the first group comprises 5 private houses on a larger residential estate. The houses differ from their neighbours only in the solar collectors installed on the southern walls. The architecture is Provençal — white walls, ridged tiles on a low pitched roof. Solar gains supply about 60% of demand (heating and domestic hot water). Air-conditioning is not included in this design.

the second group comprises three rectangular houses situated close to a new oil-fired district heating plant. The aim of this design is to supply all demand by solar energy (heating and domestic hot water).

Situation

Aramond, Gard, France.
Latitude: 43° 30' N
Temperature: −2 °C (climatic zone C)

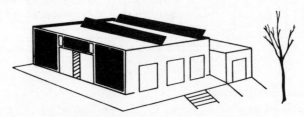

Figure 138. Aramon house Group II

Degree-days: 1,780 (1st October—31st March)

Average energy received in January: 1,114 Wh m^{-2} per day (on the horizontal plane

6.6.1 GROUP I

Building

2 storeys + basement (7 rooms), or 2 storeys (5 rooms). The 7-roomed houses are situated on sloping land directly above a road.

Energy demand: 13,884 kW for heating (October to March) 3,000 kW for hot water (October to March) (This demand information applies to the 5-roomed house)

Coefficient G

1 W m^{-3} °C^{-1} for the 5-roomed houses (which have higher insulation than an all-electric house)

G: 1.3 W m^{-3} °C^{-1} for the 7-roomed houses.

Collectors

Type: water
Position: south wall
Inclination: 90°
Area: 35 m^2 (5-roomed house), 38 m^2 (7-roomed house)
Construction: the absorber is a system of small diameter tubes, or sheets of steel stamped and welded to form a compartmental flat panel (Figure 62), or aluminium sheets joined together.
Coverplate: double glazed

Storage

Type: water
Volume: 4 m^3
Location: in basement or on ground floor
Insulation: 30 cm of vermiculite
The store is sized to compensate for the nocturnal gap in insolation, and to prevent too rapid an increase in water temperature in the collectors, which would reduce their efficiency (see earlier).

Heating

Primary circuit (collectors): water
Distribution circuit: water
Room heating: underfloor

142

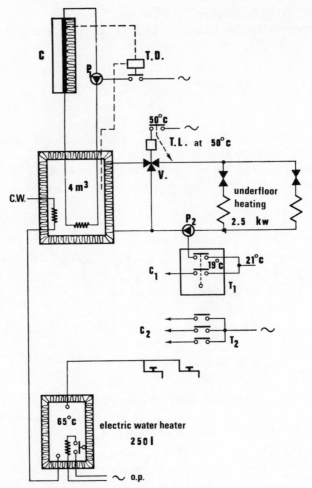

Figure 139. Heating circuit (based on a technical publication by E.D.F.)

C Collectors, C_1 electric convector in lounge, C_2 electric convectors in other rooms, CW cold water, T_1 two-stage thermostat in lounge, T_2 thermostats incorporated in the convectors, T.D. differential thermostat, P_1, P_2 pumps, T.L. water flow regulating thermostat, o.p. off-peak periods

In Section 5.2 underfloor heating is described as being well suited to the temperatures reached in the storage tank whereas water in a conventional central heating system has to circulate at 95 °C.

Underfloor heating was adopted for this scheme, and the normal distribution temperature is 30 °C.

motors: 2 circulation pumps (1 for the primary (collector) circuit, 1 for the secondary (distribution) circuit)

supplementary heating: standard electric convectors independent of the solar panels

Domestic hot water

An independent storage tank. The domestic hot water supply is preheated using an exchanger in the storage tank, and is raised to the service temperature by an electrical resistance heater.

In summer the hot water is supplied entirely by the solar panels.

Control

The control system is designed to give priority to solar energy. The room temperature thermostats have a threshold point at which they either switch on pump P_2 if the temperature falls below 19 °C or stop the pump and switch on the electric convector heaters if the room temperature has not risen after a given time.

Contact thermometers are used to control the floor surface temperature (21 °C $< t <$ 26 °C).

Another control system links the collectors and the tank as shown in Figure 139. Mechanical ventilation is used to regulate the volume of air extracted.

Performance

Measurements taken over the first few months of the experiment show that energy savings of 50% were made.

6.6.2 Group II

The object in this project was to supply all heating and domestic hot water requirements.

Building

One storey, rectangular shape, flat roof.
Area: 133 m² (12.82 x 10.39)
Height: 4 m

Insulation

Enhanced: $G = 0.93$ W m^{-3} °C^{-1}

Note: Present regulations for private housing require $G = 1.75$ W m^{-3} °C^{-1} in Zone C.

Collectors

Type: water
Orientation: south
Position: some on the south wall, some on the flat roof
Inclination: 90° (wall) and 55° (roof)
Area: 40 m²
Construction: absorber in stamped and welded sheet steel
Coverplate: double glazed

Storage

Identical to the houses in Group I

Heating

Primary circuit: water (identical to the houses in Group I)
Secondary circuit: water
House No. 1: underfloor heating in steel tubes spaced 0.15–0.35 m apart laid in 8 cm of concrete. The flooring surface is protected by polyurethane insulation. The floor temperature is designed to remain between 21° and 26 °C.
House No. 2: cast iron radiators.
House No. 3: vertical steel panel radiators.
The radiators and panels are controlled by thermostatic valves.

Conclusion

The projects above should supply more accurate statistical data than is available from extrapolations of the results for one single house, as it is well known that energy consumption can be very different for identical houses (±30%).

Another group of private houses is also under construction at Le Havre, France as part of a scheme being sponsored by the Ministère de la Qualité de la Vie.

6.7 HOUSE AT PONT-AR-BRUN, TREGASTEL, COTES DU NORD, FRANCE

Designer: J.-P. Batellier
Solar engineering: G.R.E.E.N.

6.7.1 General

This house has been designed as a second home to be occupied at different times throughout the year but later to become the main residence. The living area has been generously proportioned as the number of occupants can vary between 2 and 8 (volume to be heated: 510 m³).

(1) The following considerations resulted in the decision to use solar energy as the main source of supply for heating requirements:

the ever-increasing cost of conventional energy sources and the impossibility of orecasting future price movements,

the desire to reduce atmospheric pollution,

general mildness of the climate in this area (the temperature falls below $-2\ °C$ on only 5 days per year),

a suitable average sunshine receipt, 1,875 hours per year,

due to local meteorological conditions the cold periods (N and NE winds) correspond to a relatively high sunshine level, while rainy periods coincide with mild temperatures,

the site is open to the south, and,

the obvious benefit of maintaining the fabric of the building by being able to

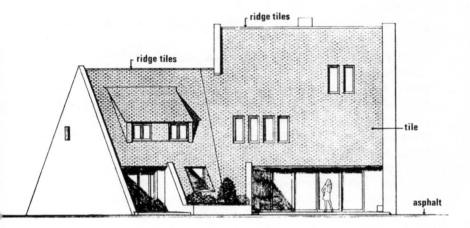

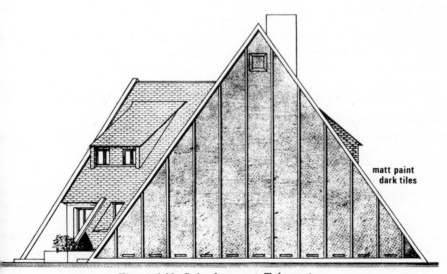

Figure 140. Solar house at Trégastel

gently heat and air the house during the periods of inoccupation with no energy expenditure.

(2) As the site is protected by local town planning regulations a relatively traditional architectural style had to be adhered to by the owner and the designer. The principle constraints were:

main view NW to SW,
cold NNE winds,
tiled roof of two equal pitches,
wall plate to be at less than 3 m,
fully-bricked gables or with the minimum of aperture.

(3) The combination of basic data and local or economic constraints led to the following design:

traditional architecture with masonry and tiled roof which continues to the ground on three façades, W., N., E.,
a blind gable which is entirely a solar collector at $S - 15°$ W,
air to be used as the heat-transfer fluid,
storage in the gable wall itself,
the requirement for forced air circulation to give an acceptable output when the house is occupied and to prevent an inversion of flow,
provision of supplementary heating by an electrical heating battery in the air circuit which would be capable of supplying the total demand after a prolonged period of practically zero solar gains.

Electricity was selected as the supplementary energy source on account of its great flexibility of use and low installation cost even though a solution using propane gas, for example, would be cheaper to run.

6.7.2 Estimation of demand

1. *Temperature*

Lowest outside temperature: $-2\,°C$,
Internal temperatures, living rooms: $20\,°C$,
                        bedrooms: $18\,°C$
                        bathroom: $22\,°C$
                        loft and garage: unheated
Air renewal:
    living rooms: 1 air change/hour
    kitchen, bathroom and W.C.: 2 to 3 air changes/hour
Additional heat losses:

	Glazing	Wall or roof
to the North	20%	10%
to the West	10%	5%
to the East	20%	10%
to the South	0%	0%

. Insulation: insulation coefficients

The building's insulation resulted from detailed study and careful consideration of the walls, roof and flooring in order to avoid thermal bridges. Double glazing was used and close attention paid to the tightness of the metal window fittings, etc.

single glazed window	$K = 5.80 \text{ W m}^{-2} \text{ }^{\circ}\text{C}^{-1}$
double glazed window	$K = 3.50 \text{ W m}^{-2} \text{ }^{\circ}\text{C}^{-1}$
external walls	$K = 0.58 \text{ W m}^{-2} \text{ }^{\circ}\text{C}^{-1}$
roofing	$K = 0.46 \text{ W m}^{-2} \text{ }^{\circ}\text{C}^{-1}$
losses through floor to ground	$K = 0.92 \text{ W m}^{-2} \text{ }^{\circ}\text{C}^{-1}$
loft ceiling	$K = 0.58 \text{ W m}^{-2} \text{ }^{\circ}\text{C}^{-1}$
internal walls to unheated areas	$K = 1.30 \text{ W m}^{-2} \text{ }^{\circ}\text{C}^{-1}$

When a 10% surcharge is included to compensate for any eventual building faults in the insulation or air tightness, the following table gives the summary of the heat demand.

	Dry Rooms		Humid Rooms	
Description	Conduction Losses (kW)	Air changes (kW)	Conduction Losses (kW)	Air changes (kW)
Living room and related volume	4.8	2.5		
Bedroom No. 1	0.67	0.20		
Bedroom No. 2	0.32	0.16		
Bedroom No. 3	0.52	0.20		
Bedroom No. 4	0.53	0.23		
Storage areas	0.32	0.08		
Kitchen			0.33	0.31
Groundfloor bathroom			0.45	0.23
Groundfloor W.C.			0.18	0.06
1st Floor bathroom			0.52	0.36
1st Floor W.C.			0.17	0.06
TOTALS	7.16	3.37	1.65	1.02
Total air changes			4.39 kW	
Total conduction losses			8.81 kW	
OVERALL TOTAL			13.2 kW	

Direct and diffuse gains through the openings have not been allowed for in the above calculation.

For a heated volume of 510 m^3 and a temperature $t = 20$ °C, the average volume coefficient is

$$G = 1.16 \text{ W m}^{-3} \text{ }^{\circ}\text{C}^{-1},$$

from which one can calculate the total demand during the heating season (1st October to the end of April) from the number of degree-days.

These are available from C.S.T.B. publications and other sources as: 2,16 degree-days, or G.V.D. 24 h, whence

1.16 x 510 x 2,164 x 24 = 30,725 kWh per year.

This theoretical value can be reduced by using a night set-back clock in th control system, which reduces the nocturnal demand by up to 5 °C.

For 14 hours of daytime pattern ($t = 20$ °C) and 10 hours of nocturnal patter ($t = 17$ °C).

(Note: Incidental gains (occupants, lighting, etc.) have to be included in th daytime pattern which bring the reference temperature for the degree-day calcu lation to $t = 18$ °C.)

Then

$$1.16 \times 10^{-3} \times 510 \times 2,164 \times 24\,h \times \frac{14}{24} = 17,923 \text{ kWh}$$

$$1.16 \times 10^{-3} \times 510 \times 1,553 \times 24\,h \times \frac{10}{24} = 9,187 \text{ kWh}$$

where 1,553 represents the degree-days for the temperature 17 °C.

Using the average temperatures collected over 30 years (see meteorological data one can calculate the monthly distribution of demand throughout the heatin season by using the number of degree-days corresponding to the two temperatur levels (20 ° and 17 °C).

	Oct.	Nov.	Dec.	Jan.	Feb.	March	April
Day pattern D	163	296	395	367	340	341	262
Night pattern D	106	206	302	274	240	250	175
Distribution (kWh/month)	1,746	3,592	4,988	4,660	4,616	4,303	3,205

6.7.3 Estimation of solar gains

Due to architectural limitations the collector gable in this project is vertical. (. better winter efficiency would be obtained by having the collectors inclined a about 15 or 20° to the vertical.) To allow for the maximum energy being receive at approximately 13.00 h (solar noon) the wall is orientated to S 15°W.

The total efficiency of the collectors has been calculated at 50%.

The inclusion of an ornamental garden in front of the wall has increased th albedo and produces an improvement of about 5% in the energy collected. Th ornamental garden falls away from the wall at about a 1:10 gradient.

The theoretical energy gains per square metre were estimated:

(a) from the atlas for total solar radiation for France (F. Tricaud's thesis and
(b) from abstracts of pyranometer measurements taken at Rennes and cor rected for the site at Pont-Ar-Brun to give the actual insolation.

If σ is the ratio of the number of sunshine hours h to the number of daylight hours H, $\sigma = h/H$; the energy received is proportional (experimental result) to $\sqrt{\sigma}$ (see also Equations 10 and 11).

Thus we have the following table:

	Oct.	Nov.	Dec.	Jan.	Feb.	March	April
/day	10.7	9	8.1	8.6	9.9	11.6	13.5
: sunshine hours/day	4.4	2.2	1.6	1.9	3.2	4.3	5.7
$= h/H$	0.41	0.24	0.20	0.22	0.32	0.37	0.42
heoretical energy received in W h^{-1} m^{-2} day^{-1}	2,838	2,134	1,415	1,671	2,445	3,270	3,206
° (after Tricaud)	3,300	2,350	1,850	2,050	2,600	3,650	3,250
.verage used in calculation (Wh m^{-2})	3,050	2,240	1,600	1,850	2,500	3,450	3,230
ractical gain in Wh m^{-2} day^{-1} with an efficiency of 0.5; albedo 1.5	1,604	1,178	842	973	1,315	1,815	1,700
.ctual collected flux on 60 m^2 useful area kWh month^{-1}	2,984	2,121	1,566	1,810	2,210	3,377	3,059
)emand in kWh month^{-1}	1,746	3,520	4,988	4,660	4,616	4,303	3,205
)eficit	–	1,399	3,422	2,850	2,406	926	146
urplus	1,238	–	–	–	–	–	–

OTAL DEFICIT OVER 7 MONTHS . .   . .   . . 11,148 kWh

In conclusion, for a demand of 27,110 kWh solar gains will supply 17,127 kWh, ır about 60 % of requirements.

The energy expenditure for supplementary electricity will be 9,983 kWh per ıeating season, or an expenditure of 1,796 F (prices for the first quarter of 1976).

In actual effect the expenditure will be lower because allowance has not been ıade for natural gains through the large panes on the west face nor has the hot air ›roduced by the log-burning fireplace been included. These natural gains can be ıstimated at 1,300 kWh.

;.7.4 Economic return on the installation

The solar gains represent 17,127 kWh or, using the value given above, 3,222 F ›er year of electricity.

The extra cost of the installation is accounted for by:

the addition of the gable wall in reinforced concrete to supply
storage .. ..   15,000 F

a glazed façade with metal framing to provide the greenhouse
effect .. ..   16,000 F

         31,000 F

The savings of the blown air system compared with a conventional
installation with radiators or convectors is approximately   ..   14,000 F

Therefore, the excess equals .. ..   17,000 F

This figure has to be compared with the annual savings on the fuel bill of
approximately 3,222 F, or 18.9% of the investment. The system, therefore, will pay
for itself over a six year period.

It should also be noted that these figures do not allow for the increasing cost of
energy nor for the considerable savings on upkeep during the periods of inoc
cupancy, when the house can be maintained at 6.7 °C above the external temper
ature at no cost whatsoever.

6.7.5 Principles and general diagram of operation (prize winner at the solar house competition, Saint-Gobain, 1976)

1. *General principle*

These installations provide central heating by hot air:

a processing plant, comprising a filter, a 1,500 m³ h⁻¹ fan and a 12 kW electric
heating battery, blows the air through a network of ducts to the rooms to be
heated,

some air is expelled from the humid rooms (kitchen, bathroom, W.C.) by
two-speed mechanical fan of 400 m³ h⁻¹ power, but the greater portion of the air
is recycled through the solar collectors on the south gable, fresh air is introduced
through a self-regulating intake on the east wall.

2. *Operation*

Three patterns of operation have been designed for:

(i) *Summer use.* The processing plant is completely cut off from the circuit, and
the installation operates by simple ventilation. The fresh air enters on the east
façade and, after passing over the collectors, is rejected at the roof through a
manually-operated shutter.

(ii) *Winter use.* Two patterns of operation have been designed for, according to
whether the house is occupied or not.

(a) when the house is unoccupied –

     the fresh air intake is closed, and

     the processing plant is by-passed by a duct with a manually-operated shutter

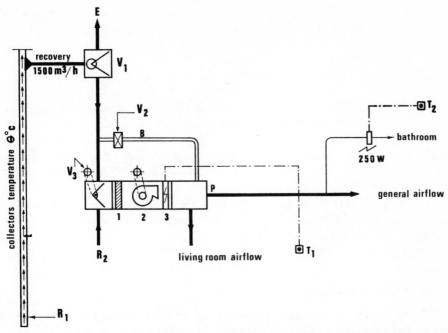

Figure 141. General diagram of the principle – blown air heating. V_1 Manually-operated shutter (winter and summer settings), V_2 manual valve, V_3 motorized shutter changing the return from the lounge if the air temperature in the collectors $< 18\,°C$. It also provides manual isolation in winter, R_2 return airflow from the lounge at $1,500\ m^3\ h^{-1}$ if $\theta < 18\,°C$, R_1 the return from the lounge to the bottom of the collectors if $\theta > 18\,°C$, P plenum for blown air system.

1. Filter, 2. fan, 3. 12 kW electric heater, T_1 room temperature thermostat – bathroom, T_2 room temperature thermostat with a night set-back clock in the lounge

The internal air is constantly circulated through the collectors and the rooms by natural convection.

In this case the efficiency of the installation is low but it is sufficient to maintain the internal temperature at 6 or 7 °C above the outside temperature with no energy expenditure. In addition, the constant movement of the slightly heated air will prevent dampness and mildew, frequent problems in unoccupied seaside houses.

(b) in the case of normal occupancy, when the exit temperature of the collectors is above 18 °C the blend of fresh air and internal air from the heated space is passed over the collectors. After filtering, it enters the air-heater which is controlled by the room temperature thermostat, and electric heaters supply any supplementary heat that is required.

When the air in the collectors is lower than 18 °C a motorized shutter short-circuits the collectors to draw the air directly from the lounge through a recovery inlet placed in the ceiling. Here again if needed, the electric heaters make up the heating deficit.

To increase general thermal comfort low voltage electric cables have been laid into the flooring (total power: 3 kW). These could ultimately be supplied by a windmill.

3. Control

The temperature of the interior is controlled by a room temperature thermostat with a night set-back clock. The thermostat controls the electric heater in the air treatment unit. To avoid introducing air from the collectors into this unit when it is colder than the ambient air, a thermostatically-controlled motorized shutter enables or disables the recycling of the internal air depending on whether the air temperature from the collectors is above or below 18 °C.

To obtain slightly higher temperatures in the bathrooms, the outlet vents are fitted with a supplementary 250 W electric heater, which is controlled by a room temperature thermostat set at 22 °C.

Every air outlet has a shutter to regulate the airflow.

4. Solar collectors and storage

The south gable of the house, approximately 70 m² in area, acts as a collector. It is formed of the following layers (from the outside inwards):
— a double wall of reinforced Plexiglass XT panels with a total thickness of 16 mm. The panels are mounted in anodized aluminium frames, retained by clips, with neoprene seals.

Plexiglass XT, which is shock-resistant and stable to a temperature of 80 °C, was selected because of its thermal insulation coefficient ($k \simeq 2.7 \text{ W m}^{-1} \text{ °C}^{-1}$), its transparency to short infrared radiation and its relative opacity to long infrared radiation (heat radiation from the wall of the storage):
— a 10 cm layer of air,
— a 40 cm thick concrete wall which serves as the thermal store. The wall is painted in the dark colour of the tiles and has a storage capacity of about 12 hours. It is independent of the main structure and can expand freely.
— a thermal insulation of 8 cm of rock wool,
— an interior load-bearing wall.

After passing through the transparent wall, the solar radiation is transformed to heat on the dark surface of the concrete wall. Some of the heat warms the air in the collectors and the balance is stored in the concrete. The stored heat is transferred to the air in the collectors during the night and during periods of low insolation, nevertheless all of this stored heat is not reusable. A small amount of it is released directly into the house and a further quantity radiates outwards through the transparent wall and is lost for good. (An improvement could be made by including an outside blind to limit the negative greenhouse effect during the night.)

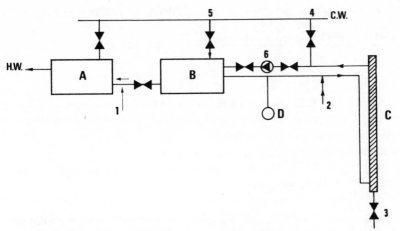

Figure 142. Diagram of domestic hot water production.

An electrically-heated store 200 l, 4 kW, B reheater–exchanger tank, 300 l, C solar collectors, 7 m² useful area, D expansion vessel,

1. preheated water, 2. antifreeze injection valve, 3. primary discharge, 4. primary replacement, 5. secondary supply, 6. pump

6.7.6 Production of domestic hot water

1. *Demand*

The demand has been calculated for three people in winter and eight in summer, with 40 l day^{-1} at 60 °C, i.e.,
in winter: $3 \times 40 \times (60 - 10) = 6.96$ kWh day^{-1}
(corresponding to 4 h of electric heating at 2 kW),
in summer: $8 \times 40 \times (60 - 15) = 16.24$ kWh day^{-1}
(corresponding to 4 h of electric heating at 4 kW).

2. *Collector area*

Winter pattern: for the five least favourable months (November to March) the practical gain per square metre of collector per day =

$$\frac{1,178 + 842 + 973 + 1,315 + 1,815}{5} = 1,225 \text{ Wh m}^{-2} \text{ day}^{-1}$$

$$= 1.2 \text{ kWh m}^{-2} \text{ day}^{-1},$$

therefore, the necessary area is

$$S = \frac{6.96}{1.2} = 5.8 \text{ m}^2 \text{ or } 6 \text{ m}^2.$$

Summer pattern: for the five most favourable months (May to September)

$$\frac{1,473 + 1,392 + 1,392 + 1,473 + 1,577}{5} = 1,461 \text{ Wh m}^{-2} \text{ day}^{-1}$$

$$= 1.46 \text{ kWh.}$$

It will be noted that there is little increase in output compared to the winter period. This is due to the vertical position of the collectors. An inclination of about 25° would, of course, be more beneficial.

The necessary area is

$$S = \frac{16.24}{1.46} = 11.12 \text{ m}^2$$

In practice only 7 m² of collectors would be used with supplementary electrical heaters supplying any unusually high demands (8 people, several weeks of the year).

3. *Installation*

The solar collectors are flat sheet metal radiators mounted in the gap between the concrete and transparent wall of the south gable.

A thermostatically-controlled pump circulates water through the collectors but disengages if the temperature of the solar heater circuit falls below 15 °C.

This circuit supplies a lagged storage tank, 300 l in size, which in turn provides preheated water to a second tank, 200 l in size, which is fitted with a 4 kW electric

In the water supply circuit for the collectors the following are fitted:

an expansion vessel,

an anti-freeze injector,

a drain tap.

7

Economic Data for Solar Energy

7.1 ANALYSIS OF COLLECTED RESULTS

Many solar projects have been undertaken in France; some have been aided by public authorities through the intermediary 'la Délégation aux Energies Nouvelles' and others have been undertaken by private individuals without public aid.

These schemes are at different stages of realization; some have been operating for several months, others are at the moment in the course of construction and several schemes have just left the planning stage. Consequently it is difficult to assess actual operating results and costs, both of which are rarely published. In fact, only the anticipated design performance figures are usually available.

7.1.1 The E.D.F. solar houses at Aramon and Le Havre

The following results relate to the group of houses already described in Chapter 6. The original intention was to achieve a savings of 50% in comparison with an identical house on the housing estate which was equipped with electric heating. To achieve a 50% saving at Le Havre in the north of France would necessitate having a larger collector area and would thus incur an even larger financial surcharge, a direct result of the less favourable climatic conditions.

The first experimental results published by E.D.F. confirm the design predictions. An installation in continuous use by also supplying domestic hot water and air-conditioning will pay for itself more quickly than an installation which only operates intermittently, for instance, providing only central heating.

At Aramon the total cost per unit of collector area was 1,657 F per square metre. One square metre was used for 3.71 m^2 of living area, and $G = 0.92$ W m^{-3} °C^{-1}. (35 m^2 of collector supplies 130 m^2 of living area in this installation.)

The surcharge was higher at Le Havre where 46 m^2 of roof-top collectors were needed for 108 m^2 of living area if the 50% savings were to be reached. For this installation the fuel savings realized are 2,000 F and it will take at least one generation to pay off the cost of installation.

7.1.2 School at Carbonne, Haute-Garonne, France

This school complex of 1,860 m² area was commissioned in September 1976. Solar energy was made feasible by reducing heat demand through an increased level of insulation and by heat recovery from the used air. The G coefficient thus drops from 1.7 W m^{-3} °C^{-1} to 0.92 Wm^{-3} °C^{-1}, which gives a 45% energy savings.

As the school is an experimental building a three year measurement programme has been planned in order to establish the energy balances and the economics of the venture. The cost of the solar installation was 280,000 F, which is expected to produce an annual savings on fuel of 7,000 F.

The solar system has 260 m² of collector, giving 1 m² of collector plate for 6.92 m² of floor area.

$$\text{Cost of solar installation per 1 m}^2 \text{ of collector} = \frac{280,000}{260}$$

$$= 1,076 \text{ F m}^{-2} \text{ of collector.}$$

The expected fuel saving is 50% but in this application the pay-back period is very long, and if the cost of the installation is compared to the anticipated savings we get

$$\frac{280,000}{7,000} = 40 \text{ years.}$$

7.1.3 Public housing in Aix-en-Provence and Avignon

Aix-en-Provence-Rousset, Bouches-du-Rhône: *'Corail'* project.

These experiments are undertaken as part of the low rent public housing programme (H.L.M) and illustrate a solar conversion to an existing scheme consisting of three 4-storey blocks with 32 flats in each.

Heating and hot water had originally been supplied by electricity ($G = 0.96$ W m^{-3} °C^{-1}), and the conversion to solar heating and domestic hot water called for 120 m² of collector to be mounted on the south gable and 126 m² on the flat roof. This gives a total area of collector of 250 m² per building.

Solar central heating is effected by hot water pipes in the flooring tile. The electric heating already installed was by individual room heaters.

1 m² of collector surface is provided for 5.4 m² of floor area.

Avignon (Vaucluse): *'La Bise'* project

This scheme is based on two 13-storey tower blocks. Initially the buildings were designed to be all-electric, with $G = 1$ W m^{-3} °C^{-1}.

In this application only domestic hot water is to be supplied by an independent solar system for each floor. The installation consists of a separate storage tank and distribution tank on each floor to service the flats.

The collectors themselves form the balustrades on the south-facing balconies.

Economic evaluation

'Corail' project (heating and domestic hot water): the savings vary from 23% in January to 100% between May and September, with the average for the year being 48%.

'La Bise' project (domestic hot water only): the savings vary from 47% in January to 100% from May to July, giving an average of 74% over the year.

Cost per square metre of installed collector

Corail project: 1,990 F (hot water + heating)

La Bise project: 1,470 F (hot water only)

The Bise installation saves approximately 11,000 F in fuel costs when compared to a night storage heating system. The investment is 317,520 F.

For 48 apartments, the installation of individual hot water tanks would cost 2,000 x 48 = 96,000 F which would bring the surcharge to 221,520.

The unadjusted pay-back period will be

$$\frac{221,520}{11,000} = 20 \text{ years.}$$

7.1.4 Conclusions

These few results demonstrate a surcharge which is only paid off after a long time, probably at the time when solar technology will be well developed. Other collector systems do cost less to instal (air or trickle-flow collectors).

7.2 THE COST OF COLLECTOR PLATES AND SOLAR INSTALLATIONS

This is purely comparative.

7.2.1 Sealed water collector

In May 1975 the retail price of a good quality collector was 603 F m^{-2} (single-glazed, without selective coating on the absorber).

The control system (2 detectors + 1 comparator) cost 1,500 F.

The complete solar water heater (250 litres), consisting of 4 m^2 of collectors with 1 circulation pump, thermostats, insulation and supplementary heaters of 1.5 kW, cost 5,960 F.

Increased production has since brought the prices detailed above down by 30%.

Some collectors use a conventional panel radiator painted black as the absorber. In this case, the price is considerably less, about 400 F.

7.2.2 Trickle-flow collectors

This is the type of collector used in the Thomason house (cf. Chapter 6). A collector similar to this was a prize-winning entry at the first H.O.T. Congress in September, 1975.

The installation of a complete heating system for a house of 100 m² costs 30,000 F, which is less than half the cost of an installation using circulating water collectors.

7.2.3 South-facing wall collector

The double glazing, aluminium framing and the connections for the ducted air system cost 250 F m⁻².

If the double wall Plexiglass XT is used at a cost of 190 F m⁻² then the total will be about 340 F m⁻².

7.2.4 The cost of a solar installation

Considering that a supplementary heater has to be supplied to compensate for those gaps in insolation which last several days, the surcharge arises chiefly from the cost of the primary circuit (collectors + storage + control) and partly from the cost of the control of the secondary circuit.

A complete installation for a dwelling of 120 m² in the South of France costs approximately 46,000 F, and produces fuel savings of 2,000 F per year. This gives a pay-back period of at least 20 years. By using trickle-flow collectors the cost would be halved.

7.3 PRESENT POSSIBILITIES

Trickle-flow collectors and storage-wall collectors could reduce the pay-back period to ten or twelve years whereas the cost of solar installations based on sealed water panels will remain high. The collectors represent only about one-third of the cost; as the remaining components (tank, control mechanism) are already in industrial production a decrease of 50% in the price of the collectors will not significantly affect the total cost.

At the present moment, if solar water heaters are used continuously they give a pay-back period varying from three to six years, according to the geographical location.

Two possibilities exist *a priori* for bringing down costs:

subsidies can be granted to industry, or grants could be made to consumers to cover some of the extra costs involved.

The second solution has been the method most often adopted to avoid creating an artificial drop in the price of the materials.

An American study has shown that customers are willing to invest approximately $2,000 for a heating installation and approximately $200 for a solar water heater.

Figure 143 gives a map of France showing solar application potentials. This map was based on the study of gains and losses on a 100 m² house, equipped with

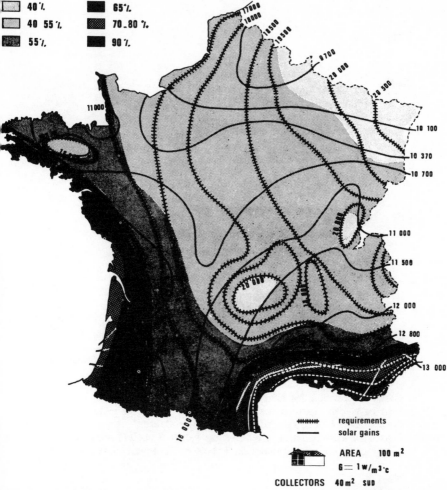

Figure 143. Map of solar energy inputs as a percentage of requirements (published by G.R.E.E.N.).

40 m² of collector on the south wall for different regions of France, central heating only.

In an economic approach to the problem of costs there is an optimal area for collectors. For example, for central heating, the collector area required on 15th February (the middle of the winter) is taken as the optimum. The deficit in collector output during January and December will be compensated for by the supplementary heating system.

E.D.F. has examined this problem for two different economic conditions, summarized in Figure 144, where it can be seen that the optimum collector area is about 40 m² when the cost is 200 F m⁻².

It should be noted that this figure is still a long way off.

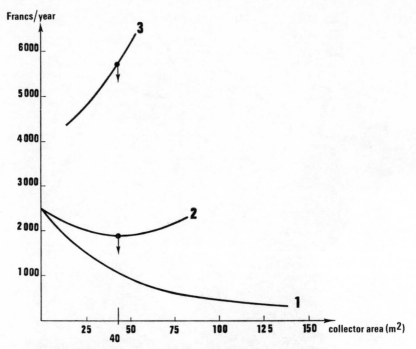

Figure 144. Optimum area of collectors (E.D.F. publication).
Annual cost of heating for a detached house of 120 m² (average French climate) as a function of size of the installation.

1. Cost of electricity for heating and domestic hot water,
2. Cost of electricity plus the annual charge on a solar installation with a price of 200 F (£25) per square metre of collector.
3. Cost with a solar installation costing 1,000 F (£125) per square metre. (The annual charge corresponds to 10% of the investment cost)

In order to allow for the annual interest and depreciation charges on the solar installation it is necessary to introduce the idea of present value of future fuel savings:

$$V = C\frac{(1 + i)n - 1}{i(1 + i)n}$$

where V = present value, C = annual heating cost, n = number of years and i = % interest rate, rather than the ratio

$$\frac{\text{total cost of installation}}{\text{annual fuel savings}}$$

which gives the number of years needed to pay for the installation without allowing for interest charges on the money and depreciation on the installation.

7.4 DEVELOPMENTS

It is by having installations (schools, swimming pools, crêches) generally public buildings, actually operating that an open attitude towards the use of solar energy will gradually be established. This attitude does not exist so far, but with the desire to preserve living standards, with the growth of medium-sized towns, with State finance for demonstration projects in several regions and with the rise in cost of other energy sources this must all lead naturally to rapid expansion in solar energy applications in the housing sector, bearing in mind that we are still in the experimental phase.

A period of active research (on the study of integrated systems under variable patterns of use and the development of economic concentration systems) will allow objectives to be more easily identified.

Financial encouragement should also be provided to people who wish to install a solar heating system in their main residence.

8
Legal Problems Posed by the Use
of Solar Energy

The use of solar energy in buildings poses several problems, some of which are detailed below*.

8.1 THE RIGHT TO THE SUN

At the present time there are two town-planning regulations specifying insolation requirements for housing.

Regulation

(1) Article 47 of the Townplanning Code (Code de l'Urbanisme), Décret 61.1298, dated 30th November 1961, official publication of 5th December 1961, states: 'When a dwelling unit comprising at least 15 apartments is to be built each building must (unless absolutely impossible) satisfy the following conditions: at least half of the façades containing openings which provide light to the living area must benefit from insolation for two hours per day on 200 days of the year. Each apartment should be so arranged that at least half of its living rooms receive daylight on façades fulfilling these conditions.

'The openings lighting the other rooms of the dwelling must not be masked by any part of a building which, at the sill of these openings, would subtend an angle of 60 ° above the horizontal plane. At least 4 m distance must be provided between two non-contiguous buildings.'

In fact this legislation only resolves one aspect of the question which is fundamental to solar technology:

(a) the proposed construction must not compromise the insolation possibilities of buildings still to be erected on neighbouring plots;

(b) and *vice versa.*

---

* Translator's Note: Some of this information applies specifically to France but is paralleled by similar legislation in other countries.

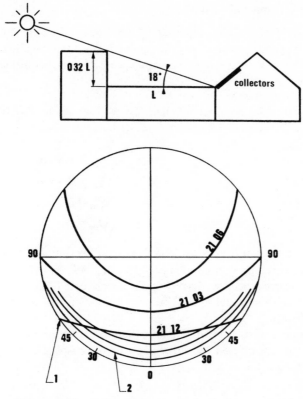

Figure 145. Solar diagram and shadows cast by buildings

An attempt, albeit limited, to tackle this wider problem is given by the 'règle du prospect' which indicates the distance to be left between two buildings as a function of their height H.

This states $L = H$ or $L = 1.5 H$.

Another form of the rule can be described as a function of the sun's altitude. In Section 2.7 solar diagrams were used to evaluate the effect of shadows cast on the collectors.

For a collector oriented to the south, the superposition of the shadow and insolation curves shows that no obstacle should have a height greater than $18°$ (below the latitude of Paris) or else a shadow may be thrown on the collector.

Nevertheless the rule of prospect does not take account of the orientation or the relative positions of the buildings with respect to the insolation. In addition, the rule only applies in the direction normal to façades and it does not specify anything for other directions.

The right to sun exists at the moment for daylight requirements, but does not include the conditions for the collection of solar energy whose raw material is free whereas all other energy raw materials cost money (petroleum, gas, etc.).

The perimeter of a supposed shadow

Just as there is a site plan determining the future development of plots, a perimeter of supposed shadow can be defined for each building in order to determine the siting of future buildings while protecting the normal functioning of solar collectors.

Defining the perimeter:

Assume a rectangular building of height H, with foundations KLMP. For each of these points one can construct a quadrilateral defined by

to the North	..	..	..	..	$N_m = 1.5 \times H$
to the South	..	..	..	..	$S_m = 0.5 \times H$
to the East	..	..	..	..	$E_m = H$
to the West	..	..	..	..	$O_m = H$

One can thus construct four quadrilaterals around the centres KLMP.

The external envelope of the points N_K, N_L N_P, etc. in Figure 146 forms the perimeter of supposed shadow. Each building constructed beyond this perimeter relative to a building A would benefit from an insolation which will not prejudice any solar installation.

The development of solar town-planning, therefore, requires more precise regulation than is currently offered by either the rule of prospect or Article 47 of the Townplanning Code.

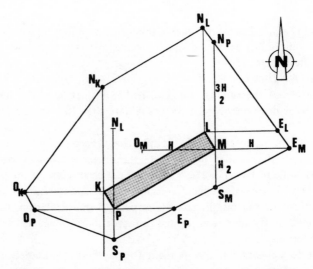

Figure 146. Defining the perimeter of supposed shadow zone, (based on M. Magnan)

8.2 EFFECT OF PRESENT REGULATIONS – BUILDING LICENCES

The problems posed by the architectural incorporation of collectors can lead to a design which will not be acceptable to the authorities who issue building licences. This would apply particularly to protected sites or to traditional architecture.

The regulations on thermal insulation (which determine the values of the co-efficient G (W m^{-3} $^{\circ}$C^{-1}) to be followed in each climatic zone and for the type of construction) are satisfied and often exceeded since the values for G are frequently lower than for electric heating.

Certain requirements at the construction stage have to be met (shutters, joints, couplings, anti-freeze, storage tank, type of piping), and these are specified in the building regulations.

The regulations for glazing materials will apply particularly in the public sector (schools). The quality of glazing used must be appropriate to its position, either in the walls or on the roof.

walls: the conventional pressure P is calculated in Pascals as $P = p(\alpha, \beta, \gamma, \delta)$ where

p = base dynamic pressure caused by the wind defined in the regulations N.V. 65–67: region I – 500 Pa; region II – 700 Pa; region III – 900 Pa,

α = site coefficient: $\alpha = 0.80$ sheltered site, $\alpha = 1.00$ normal site, $\alpha = 1.35$ exposed site, region I,

β relates to the height of the glazing above the ground:

$\beta = 1$ up to 5 storeys,

γ depends upon the frames holding the glazing:

$\gamma = 0.80$: fixed frame (the casing of the collector),

δ depends on the type of glazing:

$\delta = 1$ for non-reinforced plate glass or drawn glass,

$\delta = 2$ for laminated glass with two glass types in the layers

$\quad$ (1 Pa = 1 N m^{-2}, 10^5 Pa = 1 bar = 1 kgf cm^{-2} = 1 atmosphere).

Limitations on surface dimensions and thickness of flat glazing:

$\quad$ L – large dimension, 1 – small dimension,

$$\text{if } 1 \leqslant L/1 \leqslant 1.8 \text{ then } e = \frac{L\sqrt{P}}{10.4},$$

$$\text{if } 1.8 \leqslant L/1 \leqslant 3 \text{ then } e = \sqrt{\frac{SP}{60}}$$

(e = thickness, S = area in m^2, P = standard pressure in Pascals).

In *roofing* allowance must be made for the variation in pressure due to wind and any extra weight due to snow. At altitudes up to 200 m the normal vertical, uniformly distributed overload caused by snow has the values:

$\quad$ region I: $\quad$ 350 Pa,

$\quad$ region II: $\quad$ 450 Pa,

$\quad$ region III: $\quad$ 550 Pa.

Above 200 m the loads increase with altitude, thus in mountainous regions a correction must be made which should appear in the listing of loads.

Example: reinforced glass in a roof

	Maximum admissible span for	
Snow Pressure	6 mm thickness	8 mm thickness
350 Pa	74 cm	98 cm
400 Pa	71 cm	94 cm
500 Pa	65 cm	88 cm
600 Pa	61 cm	82 cm
⋮	⋮	⋮
2,100 Pa	35 cm	48 cm

Regulations concerning roofs

The optimum inclination of the collectors can be a constraint upon their incorporation in a design. According to the roofing material (slates, tiles, shingles, flat slabs) there is a minimum pitch which has to be respected for water run-off and snow.

This may be restricted by local town-planning regulations (regarding roofs, integration with the site). The Building Regulations Handbook supplies the particulars needed when designing roofs.

8.3 THE RESPONSIBILITY OF DESIGNERS AND INSTALLERS

The design of an installation is binding upon its author, and the law is the same as that normally applied in the building industry where responsibility is for 10 years.

Nevertheless, as the installation is designed for a *typical* year there is a non-predictable element which must be considered in any contracts. Installers are subject to the same obligations as would apply for any heating installation and the guarantee on materials is generally two years.

8.4 SOLAR ENGINEERING MATERIALS AND THEIR PERFORMANCE

A study of the performance of solar collectors currently on the market has been initiated. The trials are to be performed by CETIAT, Lyons, who will test the efficiency of the heaters and radiators. However, this body will only supply the results of the trials to the manufacturer.

At the present time several other bodies are carrying out tests on collectors: E.D.F. Les Renardières, C.E.A., and le COSTIC.

Some of the performances are published in the form shown by Figure 59, the measurements having been made with pyranometers of Class Z.

The manufacturers usually supply this type of information in their brochures.

9

Future Developments

In this book we have tried to demonstrate that there is an immediate economy in using solar energy for the supply of some of a dwelling's heating and for its air-conditioning. This is especially true for regions where there are large temperature variations during the day or between the seasons as the economics of an installation are greatly improved by using the solar energy for both heating and cooling the home and its attachments.

The processes detailed have all been tested over several years — indeed for some of them, over several decades. If research and development in this field were to blossom the advances made in several years' time would transform daily life through the modification and diversification of locally available energy sources. The spread of solar energy applications and the resultant variety of energy sources will lead to a more efficient management of the ecosystem.

What then are the main paths of solar energy research, and which of these paths will lead to medium term results?

9.1 ENERGY STORAGE

The possibilities for energy storage are very limited at the present time and condemn solar energy to being used as a 'top-up' source of supply. This sentence has been passed because it is impossible to store the energy received in the summer for winter use. However, several possible solutions look hopeful for definite progress on the problem of storage.

Those systems which call for high temperatures to produce a suitable output will use solar 'furnaces' similar to that used at the National Centre for Scientific Research (C.N.R.S.) at Odeillo, Font-Romeu. The solar furnace there is the basis of a process for a solar electric power station, and has been used for ten years in high temperature metallurgy work.

The high temperatures which have been obtained at Odeillo have led to research into the direct thermal decomposition of water at a temperature over 3,000 °C. But, even if present research leads one to think that engines will use hydrogen as a fuel, it is out of the question that these methods could be applied in the housing

167

sector let alone in private housing. It is most probable that hydrogen will be 'produced' industrially and distributed in a similar way to conventional fuel. Nevertheless, at the moment it is impossible to visualize the short-term development of private equipment for the production of hydrogen from solar energy sources.

Only 10% of the total energy emitted by the sun is found in the ultraviolet band of the spectrum (see graph, Figure 5). Much of this energy does not reach the earth's surface but is absorbed and scattered in the upper atmosphere. Furthermore, animal and vegetable life are only possible in an environment which filters out this solar energy. Photochemistry is based on reactions using ultraviolet radiation. The irreversible photochemical reactions, which create organic fuels, and the reversible ones (energy storage) can only use solar energy if the photosensors allow the transfer of energy to the shorter wavelengths.

Other methods of storage involving the change of states of molten salts do not seem suitable only for solar generating stations but would appear to have housing applications since the storage volumes required would be small.

9.2 PHOTOSYNTHESIS – PHOTOBIOLOGY

The problem of world hunger can be reduced to an energy problem. Every industrialized agricultural improvement is the result of an injection of energy, not only through mechanization but due to the manufacture of artificial fertilizers, particularly nitrates. Much research has been done on the biological fixation of nitrogen but the production of hydrogen by photosynthesis and the manufacture of certain fuels by fermentation look more promising than research into the raising of the efficiency of natural photosynthesis itself. In practice, bioenergetics has demonstrated that the maximum theoretical efficiency cannot exceed 6%. In tropical regions the efficiency of photosynthesis reaches 1.5% as compared to 0.6% in Northern France. It is pointless to hope for a large improvement on this figure, which limits agricultural production. As well as the redeployment of agriculture which is made possible by these new techniques, such essential problems as the storage of produce in dry surroundings, the ventilation of premises and the drying of produce can be solved right now with techniques based on the same concepts used for solar dwellings.

9.3 PHOTOCELLS

Solar cells have proved their reliability since the beginning of the space programme, but it must not be forgotten that the solar cells used on satellites produce less energy than was required for the manufacture of the solar cell. It is possible to build a house in a desert where solar energy would supply all its energy needs (air-conditioning, domestic hot water, heating, and electricity) through a combination of solar cells and thermal storage tanks. However, the cost would be in six figures! (NOTE: Solar cells cost about £20 per watt.) Nevertheless, this technique is viable for small requirements (of about 50 W) such as in self-contained

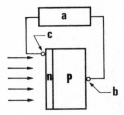

Figure 147. Diagram of a silicon photocell.

a. Load,   b. ohmic contact

radio buoys in navigation networks. However, it is probable that rapid industrial development of solar cells will soon lead to self-contained electricity installations. In fact the techniques for the production of electricity from semiconductor photocells is well known and it is not necessary to develop it here.

Three approaches are currently being developed. The first uses silicon, of which there are unlimited supplies, whereas the other two depend upon resources which are known to be limited, cadium sulphide and cadium arsenide. The projected cost for these techniques has led the Energy Research and Development Authority in the U.S.A. to estimate that by 1985 photocells will be available at £0.30 per watt. This means that those people who assured for themselves partial autonomy in heating and air-conditioning in 1975 through solar energy will be able to complete their installations to meet *all* their daily electrical requirements in 15 years time.

If solar photocells are able to supply the electricity necessary for individual installations then the need for electrical and mechanical energy will bring about the development of equipment based on both flat plate and focusing collectors.

9.4 SOLAR GENERATING STATIONS

Solar generating stations, several examples of which are in the course of construction at the moment, can be classed under three headings.

(a) *Solar power stations using a point focus* (see Figure 148). The very high cost

Figure 148. Diagram of a solar power station
(from an American publication)

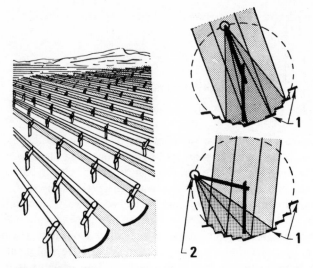

Figure 149. Diagram of a solar power station using a
line focus

of the prototypes can be explained by the fact that the solar energy is collected and concentrated on a boiler at the top of a tower. This implies a scale which is appropriate only to small towns in industrialized countries. The plane mirrors must be able to follow the movement of the sun and the precision of the mechanical equipment limits the possible size of such installations. For a 10 megawatt power station, the furthermost mirrors must follow the sun with an accuracy of 0.5 minutes. It would, therefore, be very costly to increase the collecting surface by a factor of 4 or 10, and at the present time it is illusory to imagine that the power of such stations could exceed 10 megawatts. The energy concentrated at the top of the tower drives a steam turbine just as in the old thermal power stations. Heat transfer liquids other than water will probably be used but the principle will remain the same.

(b) A different approach lies in the modular technique of *power stations using fixed mirrors and a line focus*. As shown in Figure 149, such a technique is more suitable for countries with low population densities. It is a response to the necessity of replacing diesel engines with equipment of the same capacity, in locations far from power stations. Less precise technology can be employed with this method since the mirrors do not have to follow the movement of the sun. Only the heat transfer pipe has to be moved, following a programme which can be calculated a year in advance.

(c) The third process is based on *'solar ponds'*. These ponds contain a variable density liquid which stores the solar energy.

In another application field, low efficiency solar engines are in service at the present time. They replace 10 hp diesel engines and thus provide the energy needed to run an air-conditioning installation which requires a minimum energy to operate (for the pumps, valves, etc.). They give an even greater economy of operation to dwellings which already run partially on solar energy.

9.5 CONCLUSION

The use of solar energy for heating and supplying hot water to a dwelling is already possible. Economic conditions at the moment are such that the extra investment required will be recouped in a minimum of ten years, but certain applications do give economical and dependable service in meeting the essential requirements for a house.

The development of solar technology is justified and reinforced by the desire for national energy sufficiency, but solar energy seems best suited to answering the problems of energy supply at the family, rather than the national level, or, at best, being used to supply several families.

However, modern society concentrates people into larger groupings than families, and this poses problems which are altogether different.

One field of research concerns itself with studying the advantages and disadvantages of using solar energy at the town level. All the energy-flows into a town and out of it would be considered with the aim of reaching a solution. In this way it might be possible to find different answers to the problems facing us.

Index

172